2nd International Conference on Science and Technology (CIWT 2017)

Current Issues in Water Distribution and Treatment

E3S Web of Conferences Volume 59 (2018)

Brenna, Poland
31 May – 2 June 2017

Editor:

I. Zimoch

ISBN: 978-1-5108-7316-2

Printed from e-media with permission by:

Curran Associates, Inc.
57 Morehouse Lane
Red Hook, NY 12571

Some format issues inherent in the e-media version may also appear in this print version.

Printed by Curran Associates, Inc. (2018)

For additional information, please contact EDP Sciences – Web of Conferences
at the address below.

EDP Sciences – Web of Conferences
17, Avenue du Hoggar
Parc d'Activité de Courtabœuf
BP 112
F-91944 Les Ulis Cedex A
France

Phone: +33 (0) 1 69 18 75 75
Fax: +33 (0) 1 69 28 84 91

contact-edps@webofconferences.org

Additional copies of this publication are available from:

Curran Associates, Inc.
57 Morehouse Lane
Red Hook, NY 12571 USA
Phone: 845-758-0400
Fax: 845-758-2633
Email: curran@proceedings.com
Web: www.proceedings.com

TABLE OF CONTENTS

2nd INTERNATIONAL CONFERENCE ON SCIENCE AND TECHNOLOGY

Organised by:

Silesian University of Technology

Faculty of Energy and Environmental Engineering

Current issues in water treatment and water distribution

31 May- 2June 2017

Scientific patronage:

Committee for Civil Engineering and Hydroengineering of Polish Academy of Sciences
Section of Sanitary Engineering

Scientific Committee
Chairmen: *Associate Professor Izabela ZIMOCH*

WATER – *one cannot say*
that you are necessary in life – you are life itself
Antoine de Saint-Exupéry

INTRODUCTION

Water is life, water is invaluable treasure. Therefore, today – at the time of intense growth of civilisation, and so in the face of global threat of water deficit – it is well worth stopping and thinking what the universal right to water means. Is it merely a catchphrase or the actual obligation of each country, or maybe a commitment towards the inhabitants of Africa and Asia in terms of search for safe water? The historical resolution 64/292, setting forth the "access to safe drinking water and sanitation" as a fundamental human right, was adopted the UN General Assembly in July 2010 by support of 122 countries as a result of endeavours of Pablo Solón Romero, UN ambassador of Bolivia, among others. He referred to its role in the human life and its meaning for the growth of contemporary economy and he presented moving facts about a constantly growing number of people dying of lack of access to clean drinking water. Romero emphasised the fact that it is shameful to remain indifferent to ensuring universal drinking water supply when in developing countries a child dies of water-induced diseases every 3.5 seconds.

Facing our everyday challenges, we fail to ponder over the issue of the live-giving nature and strength of water. No one disputes the fact that water is one of the most important resources on Earth. Moreover, we also fail to notice that the growth in water demand results not only from a growing global population and changing consumption patterns, but also from growing needs of dynamically developing economy. For instance, to produce 1 kg of beef, one needs ca. 15 thousand litres of water, one cup of coffee requires as much as 140 litres and a cup of tea – 30 litres. Therefore, our every-day choices have a practical effect on the use of water resources and they form the most significant needs of contemporary economy development.

Those needs result in the fact that the universal access to clean drinking water which does not pose threat to health has become a global goal, the achievement of which is ensured by the synergy of activities revolving around the sustainable consumption of water resources. Such plotted activities of the water management help support the development of innovative solutions in terms of water supply and create market opportunities for those innovations. That water will be supplied in high amounts is ensured by highly effective technologies of treatment and distribution, which reduce water intake from the natural environment, as well as procedures optimising hydraulic parameters of the system operation. Those activities are to result in decreasing the energy consumption of water supply system use and minimising the costs of water supply to the recipient. The performance of those activities is a challenge for the scientific world, dictating the necessity of close cooperation with business entities.

The aim of *„Current issues in water treatment and water distribution"* is to present the latest results of studies conducted in this respect in Poland and in the world. It includes 30 studies that cover the issues connected with modern technologies of intake, treatment and distribution of water and with the management of water-pipe networks.

I hope that this edition of research articles will be an inspiration for new challenges in the area of water supply systems.

Proceedings Editor
Izabela Zimoch, Prof. Ph.D., D.S.

Statement of Peer review

In submitting conference proceedings to *Web of Conferences*, the editors of the proceedings certify to the Publisher that

1. They adhere to its **Policy on Publishing Integrity** in order to safeguard good scientific practice in publishing.
2. All articles have been subjected to peer review administered by the proceedings editors.
3. Reviews have been conducted by expert referees, who have been requested to provide unbiased and constructive comments aimed, whenever possible, at improving the work.
4. Proceedings editors have taken all reasonable steps to ensure the quality of the materials they publish and their decision to accept or reject a paper for publication has been based only on the merits of the work and the relevance to the journal.

Title, date and place of the conference

2 nd International Conference on Science and Technology
Current Issues in Water Treatment and Water Distribution
May 31 to June 2 2017 Brenna

Proceedings editor(s):

Izabela Zimoch

28.09.2018 r.

Date and editor's signature

17, avenue du Hoggar - PA de Courtabœuf – BP 112 - 91944 Les Ulis Cedex A (France)
Tél. : 33 (0)1 69 18 75 75 - Fax : 33(0)1 69 07 45 17 – www.edpsciences.org

Membrane surface morphology and fouling in filtration of high DOC water

Beata Gorczyca[1,*]

[1] University of Manitoba, Department of Civil Engineering, Faculty of Engineering Winnipeg, 15 Gillson St., MB R3T 5V6 Canada

Abstract. In Canada many potable water sources contain very high concentrations of Dissolved Organic Carbon (DOC), accompanied by a wide range of hardness. DOC reacts with chlorine used in water disinfection to form potentially carcinogenic chlorine disinfection by-products – Trihalomethanes (THMs) and haloacetic acids (HAAs). Dual membrane plants that combine microfiltration (MF) and ultrafiltration (UF)or nanofiltration (NF) can remove DOC and reduce THMs concentration, but these plants are prone to serious fouling of their UF or NF membranes. The objectives of our research are to study the mechanisms of UF/NF membrane fouling. We have determined various resistances of DOW Filmtech NF90 (flat sheet coupon), based on the resistance in series model. The experiments were conducted on a bench scale cross-flow membrane filtration unit (Sterlitech), using synthetic water with DOC of 11 mg/L and calcium hardness of 350 mg/L that represents typical surface waters in Manitoba (Canada). The results suggest that gel layer on the surface of the membrane has a significant contribution to the flux decline. Atomic Force Microscope (AFM) allowed for relatively inexpensive, non-destructive analysis of the surface area of the gel layer deposited on the membrane filter. The morphology of the gel layer was related to the gel layer resistance.

1 Introduction

In Canada many potable water sources contain extremely high concentrations of Dissolved Organic Carbon (DOC), often exceeding 20 mg/L. DOC reacts with chlorine used in water disinfection to form potentially carcinogenic chlorine disinfection by-products – Trihalomethanes (THMs) and haloacetic acids (HAAs). The high water DOC is the primary reason why ~70% of the potable water systems in the Province of Manitoba (Canada) exceed Federal and Provincial water quality standards for THMs of 100 ppb [1].

Dual membrane plants that combine microfiltration (MF) and ultrafiltration (UF) or nanofiltration (NF) can remove DOC and reduce THMs concentration, but these plants are prone to serious fouling of their UF/NF membranes.

In the Canadian Prairie region, the water DOC can range from 11-20 mg/L and water hardness can range from 50 mg/L-350 $CaCO_3$. The DOC and divalent cations (water hardness) can combine to promote rapid fouling and reduce UF/NF membrane life, leading to reduced performance and increased operation costs for the water treatment plants in this region.

American Membrane Technology Association (AMTA) recommends that to control fouling, water DOC must be reduced to below 2 mg/L prior to NF [2]. Our research indicates that, at most, 70% of DOC can be removed from Canadian Prairie waters, which is insufficient to bring it down to 2 mg/ [3-8]. Also, the reduction of one water quality parameter, namely DOC, does not guarantee prevention of fouling. This is because fouling depends on many water quality parameters and their interactions, specifically the interactions of DOC with metal ions, as well as membrane properties and operations. Therefore, the specific mechanism of fouling needs to be understood to control fouling of membranes operating on high DOC waters.

The highest natural water DOC currently reported in membrane filtration studies is 13 mg/L which is at the low end of the concentrations observed in many Canadian surface waters supplies [9-11].

Only literature dealing with wastewater treatment plant effluents has looked at membrane filtration of water with DOC higher than 13 mg/L. Therefore, one of the aims of this research is to address this knowledge gap by investigating the membrane fouling mechanisms in filtration of natural waters with extremely high DOC concentrations accompanied by high water hardness. Although the characteristics of wastewater effluents differ from natural waters; it may be possible, that some mechanisms of membrane fouling reported for membrane bioreactors may be applicable to organic rich natural waters [12-15].

* Corresponding author: Beata.Gorczyca@umanitoba.ca

1.1 Mechanism of membrane fouling

1.1.1 Membrane resistance

In membrane filtration the water flux decline is caused by a decreased driving force (transmembrane pressure) and/or an increased membrane resistance. Generally, several mechanisms have been postulated to explain why membrane resistance increases during filtration, including: a) loss of porosity due to membrane pore blockage b) formation of the fouling (gel) layer, and c) changes to the fouling layer structure [16,17-21]. Although the gel layer can usually be removed by physical cleaning, this type of resistance to filtration can sometimes represent more than 90 % of the total resistance for membrane filtration [22].

For MF and ultrafiltration membranes (UF) the modelling of membrane resistance to filtration is often based on classic filtration theory that includes Stokes', Darcy's and Carman-Kozeny's Laws [16, 22]. Although the transport of water through NF and some UF membranes is modelled typically using the thin-film model, it has been reported that combination of these two modelling approaches gives the best results [23].
According to filtration theory, for low feed concentrations and low transmembrane pressures, applicable to natural water filtration, the permeate flux can be modeled using the modified Hagen-Poiseuille equation [24]:

$$J = \frac{e\, dp^2 (P_T)}{32\, x \mu} \tag{1}$$

where:
J - permeate flux (volume permeate/(membrane area x time, L m^{-2} h^{-1})
e - membrane surface porosity
dp - mean pore diameter
P_T - net transmembrane pressure = (P_F-P_P)
P_F - feed pressure, P_P = permeate back pressure
X - pore channel length
μ - absolute viscosity of permeate.
The equation (1) can be simplified to Darcy's equation:

$$J = \frac{P_T}{(R_t)\cdot\mu} [\mathrm{L} \cdot m^{-2}\, h^{-1}] \tag{2}$$

where:
R_t - the total resistance (m^{-1})

The resistances which can occur during a filtration process are membrane resistance (R_m), fouling resistance (R_f - consisting of reversible and irreversible fouling) concentration polarization resistance (R_p), and gel layer resistance (R_g) (Fig. 1). Assuming a resistance-in-series model, the total resistance is equal to the sum of all other resistances:

$$R_t = R_m + R_f + R_p + R_g \tag{3}$$

Determination of various membrane resistances using equation 2 requires measurements of the flux at different filtration pressure applied. The estimate of different membrane resistance can indicate the contribution of different mechanisms to total membrane fouling; however, it provides little information on the pattern of the events that leads to fouling. Eq. 2 as opposed to Eq. 1 does not contain information on membrane parameters directly related to mass transfer through the membrane such as porosity, pore diameters and pore channel length. In classical filtration theory these parameters are directly related to the mechanism of membrane fouling; but they can only be determined through by conducting a membrane autopsy.

The thin-film model is derived by equating the connective flux of the solute towards the membrane surface with the diffusive flux of the solute away from the membrane surface. Assuming this model, the flux through the membrane can be represented by [24]:

$$J = kln(\frac{Cgel}{Cretentate}) \tag{4}$$

where:
k - mass transfer coefficient (same units as flux J)
C - concentration of the solute.

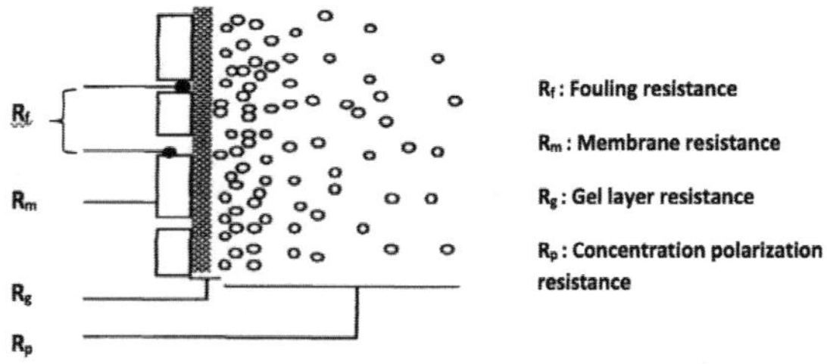

Fig. 1. Possible resistances against solvent transport through membranes [25].

1.1.2 Autopsy of membranes used in filtration of potable water sources

Properties of the virgin and fouled membrane can be studied through membrane autopsy. Current membrane autopsy protocols are based on methods used for foulants encountered within the food industry. Foulant analyses typically consist of: quantification of foulant dry density in mg/cm^2, determination of foulant composition, microbiological testing, and investigation of the foulant structure with Scanning Electron Microscopy. Most of these analyses are conducted on dried, fouled membrane. Yet, natural water membrane fouling layers, similarly to biofilms, are often composed of organic materials with extremely low density that can occupy significant volume, such as soluble microbial products [26]. Structures of such materials can change during drying, therefore, it can only be investigated through non-destructive, wet state analysis. Such analysis require the use of Confocal Scanning Laser Microscope (CSLM) or Small Angle Neutron Scattering (SANS) methods which are extremely expensive [17-21]. Atomic Force

Microscope (AFM) allows for relatively inexpensive analysis, non-destructive analysis of the surface area of the gel layer deposited on the membrane filter.

1.1.3 Relationship between external surface and internal structure of membrane fouling layer

Several techniques applied to study external structures of particle aggregates (flocs) and biofilms can be used, with modifications, to study the same properties in membranes. Biofilms can be represented as a group of compressed flocs immobilized on a substratum and several similarities between flocs and biofilms have been reported in the literature. It is difficult to describe the floc or biofilm structure using concepts of conventional geometry, due to "chaotic" distribution of particles inside these biomasses; however, floc and biofilm structure can be described in terms of fractal geometry [26-29].

Knowledge of the internal structure of the membrane fouling gel layer can provide information on the pattern of events that lead to its formation.

External surface morphology is closely related to the internal structure. For example, Fig. 2 shows that surface morphology of the gel layer formed by deposition of spherical particles will be quite different that that formed by deposition of fractal object (snow flakes in Fig. 2).

For example, the surface roughness, measured as the height difference between the peaks and valleys will be the same for the fouling layer formed with spherical particles. For deposits formed by the „snowflakes" the surface roughness will depend on the magnification (resolution) at which the measurements is conducted. Our earlier research on fractal structures of flocs suggest that deposit shown in Fig. 2 b may have lower resistance (higher permeability) than the Fig. 2 a [26-29].

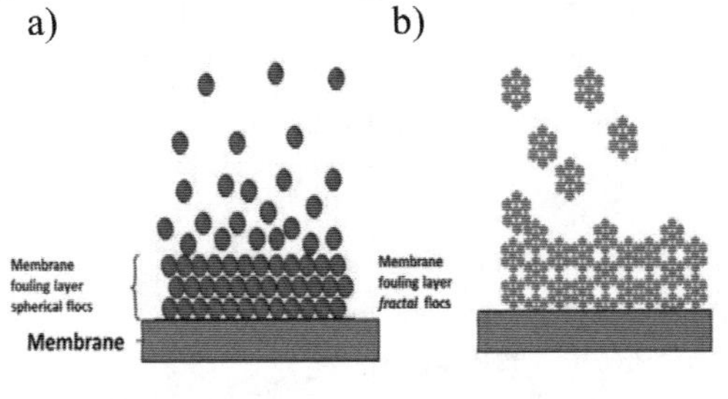

Fig. 2. Possible relationship between the structure of flocs and surface morphology of the membrane fouling layer.

The goal of our research is to obtain information on the mechanisms of fouling of UF/NF membranes used to filter high DOC and hardness surface waters. The specific contribution of this work is in linking the measurements of membrane resistance with the observation of changes in membrane surface morphology.

The long term goal of this research is to extend the life of membranes used to filter challenging surface waters, by understanding the fouling process.

2 Materials and methods

Morris Regional Water Treatment Plant (WTP), located in the Town of Morris (Manitoba, Canada), was upgraded in 2009 and officially opened in 2010 as a dual membrane treatment process. The upgrade converted a conventional lime/soda softening process to an integrated membrane treatment plant (microfiltration followed by nanofiltration). The schematic diagram of the water treatment plant in Morris is shown in Fig. 3.

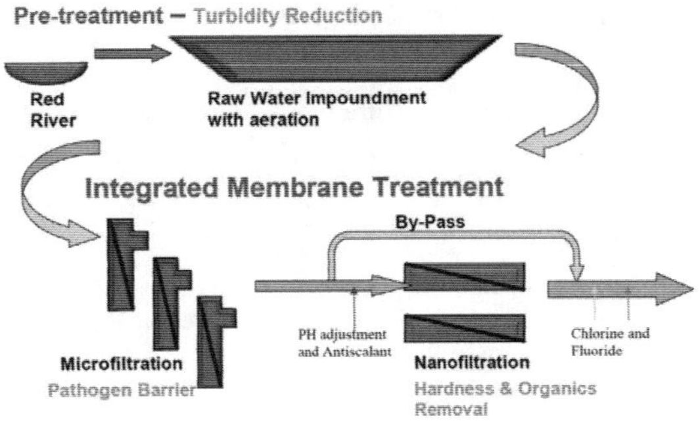

Fig. 3. Morris Regional Water Treatment Plant [30].

The plant operation is fully automated and runs 24 hours a day in summer (two nanofiltration skids run on alternate days) and 18-20 hours a day in winter depending on the demand at a constant present flow rate with a predetermined permeate flow. The membranes (Table 1) are periodically cleaned manually in place (every 2 – 3 months) using a solution of sodium hydroxide (NaOH) and citric acid as cleaning agents.

Table 1. Properties and of membranes used in the plant and the laboratory experiments.

Membrane Properties	Type of Membrane	
	TMG20-400C	NF90-400/34i
Membrane Material	Cross Linked Fully Aromatic Polyamide Composite	Polyamide Thin Film Composite
Molecular Weight Cut-off (Dalton)	< 200	90 - 200
Maximum Pressure (bar)	25	41
Maximum Temperature (°C)	45	45
NaCl Rejection (%)	99	85 – 95
Cross Flow Velocity (m/s)	0.374	0.34

The cleaning procedure is effective in restoring the permeability of the membranes and cleaning any reversible fouling occurred at the unit. However, due to high level of irreversible fouling, the NF membranes need to be replaced prematurely, which makes the treatment system uneconomical.
The NF membranes used in the water treatment plant are ultra low pressure Toray element for brackish water (TMG20-400C, Toray). The membrane specifications and operating conditions as specified by the manufacturer are presented in the Table 1. Fresh coupon of TMG20 using synthetic water was not available for the laboratory testing and a similar membrane NF90 DOW Filmtech was used (Table 1).

2.1 Laboratory determination of membrane resistances

Bench-top membrane filtration system was used to determine resistance of NF membrane. Feed water was pumped through a WEG Hydra-Cell constant flow pump forcing 7 L/min resulting in cross-flow velocity of 3.36 m/s. Water was pumped into a Sterlitech CF042D membrane filtration cell and ran in crossflow through the encased membrane coupon. Temperature of the water was carefully controlled through the heat exchanger. A flow diagram of the experimental setup can be seen in Fig. 4.

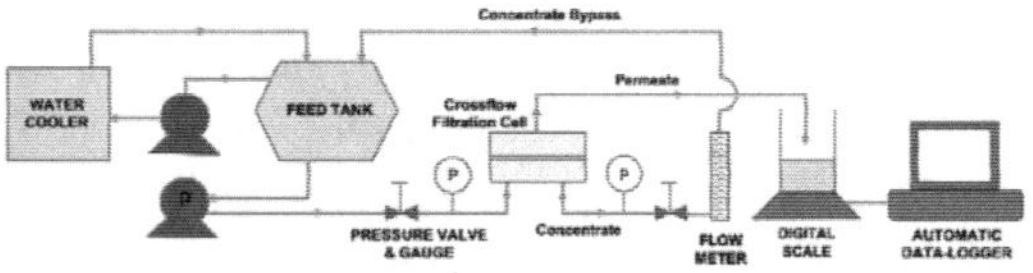

Fig. 4. Bench scale laboratory membrane filtration set-up [31].

The various membrane resistances were estimated using the resistance in series model (eq. 1 and 2). Synthetic water (Table 2) used in bench scale laboratory testing was prepared to mimic the DOC and hardness composition of the Red River water (Table 3).

The foulant accumulated on NF filters used to filter the Red River water (Manitoba) was reported to be composed primarily of hydrophilic neutral NOM fraction [31]. Based on that analysis, sodium alginate -an anionic polysachharide, which is a hydrophilic neutral compound was used to mimic natural water DOC. Calcium chloride, was used to impart hardness to the synthetic water. The concentration of the sodium alginate and calcium chloride used in the synthetic water was based on the concentration of Dissolved Organic Carbon (DOC) and Calcium hardness in Red River, as reported by Morris Plant (Table 3). NF90 membrane was compacted for 10 hours and fouled with synthetic water for 72 hours.

2.2 Membrane Autopsy

Gel layer external surface. The surface morphology of the foulant formed in the laboratory testing was observed using Atomic Force Microscope (AFM).

Table 2. Composition of synthetic water used in determination of membrane resistance.

Constituent	Formula	Molecular Weight	Concentration
Sigma Aldrich – Sodium Alginate	$(C_6H_7NaO_6)_n$	198 g/mol	11.6 mg/L as DOC
Calcium Chloride	$CaCl_2$	110.98 g/mol	350 mg/L as $CaCO_3$

Table 3. Quality of raw and treated water at the Morris Water Treatment Plant [30].

Parameter	Red River Water
pH	8.23
Alkalinity, Total (as $CaCO_3$), mg/L	261
Dissolved Organic Carbon (DOC), mg/L	11.6
Total Hardness, mg/L as $CaCO_3$	350
Calcium, mg/L	70.4
Magnesium, mg/L	42.2

A NanoScope IVa Controller (Veeco Dimension 3100) was used to show the change in foulant surface roughness during filtration. A 100μm^2 surface area was plotted in contact mode at a speed of 10μm/s. Three-dimensional surface images were collected for analysis.
Gel layer thickness. It was not possible to measure the thickness of the gel layer as the gelatinous foulant formed by synthetic water was difficult to section. However, the depth of the gel layer was measured on the fouled membrane colected from the Morris plant. Altough this membrane was fouled for much longer time that the coupons in the laboratory, it could provide some information on the relative thickness of the gel layer in comparison to the total membrane thickness. The membrane samples were carefully sectioned using sharp razor blade in wet condition and the best sections were used to view using Environmental Scanning Electron Microscope.

3 Results and discussion

3.1 Membrane resistance

The various resistances of the NF90 membrane determined in the laboratory scale apparatus are shown in Table 4. Total resistance of the TMG20 membrane collected from the plant was estimated in the laboratory to be 4.31E+14 m^{-1} and was comparable with the total resistance obtained for NF90 membrane. The difference between the values should be mainly due to the much longer filtration time applied at the WTP. The similarity of the total resistance of the membrane used in the laboratory study to the one collected from the plant validates the preparation of the synthetic water. The resistance results indicate that the fouling (R_f) and gel resistance (R_g) are primarily responsible for the flux decline of the tested NF membrane. Analyses of the gel layer external surface was conducted to obtain more information on the pattern of event that have led to its formation.

Table 1. Various membrane resistances as determined in the laboratory.

Feed Water	Rm (10^{14} m^{-1})	Rt	Rp	Rg	Rf
Synthetic	1.17	3.88	0.55	0.97	1.19

3.2 Membrane Autopsy

3.2.1 Analyses of membrane surface morphology from Atomic Force Microscope (AFM) images.

In our earlier research, we have observed vastly different surface morphologies of in a number of DOW Filmtech NF membranes using AFM [15]. These membranes have different resistance which could be related to their different internal structures of the membrane and foulant, which in turn can be reflected in different surface morphology.

Changes in the virgin membrane surface morphology are certainly determined by the mechanism of fouling. The surface morphology of the two membrane gel layers shown in Fig. 5 is certainly different. The fouled membrane, which has higher resistance to flux, has markedly higher peaks than the clean membrane.

Fractal dimensions can provide information on the pattern of events leading to development of fouling layer. AFM profiles of NF90 virgin and fouled membrane are currently being analyzed using the box counting method [33].

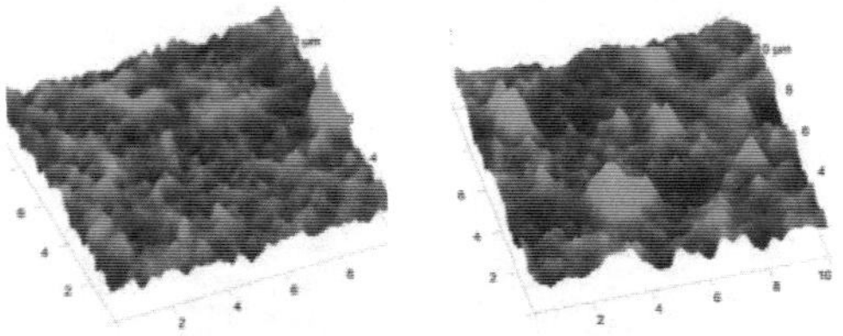

Fig. 5. DOW Filmtech NF90 membranes surface morphologies (left –clean, right-fouled).

3.2.2 Estimation of thickness of the gel layer

The images of cross-sections of the membrane collected and the plant is shown in Fig. 6. The cross-sectional images were very clearly identifying the smooth surface and depth of the gel layer on the membrane surface. Some cracks were visible on the top surface of the membrane samples, which might be a result of membranes drying and organic foulant flaking off. The estimated depth of the gel layer was about 27 μm and represented approximately 25% of the total thickness of the membrane.

Scale of the ESEM Image: 1 cm = 19.23 μm. Depth of Foulant (measured at three consecutive points) = 26.92 μm. Compacted Membrane Thickness = 107.69 μm.

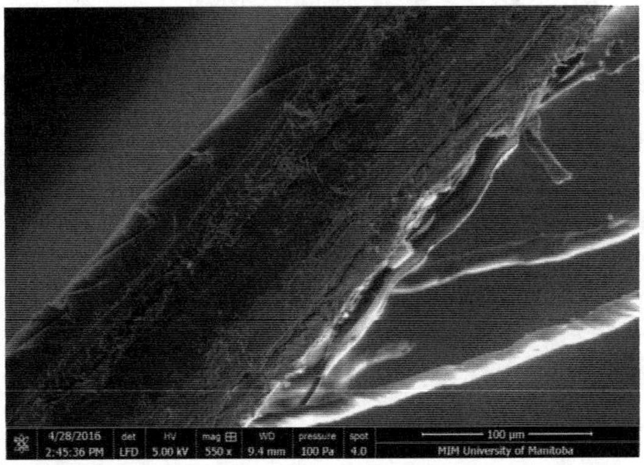

Fig. 6. Cross-section of nanofilter (TMG20) collected from Morris WTP [32].

4 Conclusions

The main operating cost of membrane filtration technologies lie in the fouling problem. In this research we have been investigating the mechanisms of fouling of NF membranes specific for filtration of challenging waters containing high concentration of DOC and carbonate hardness. The objective was to understand the mechanism of fouling to be able to control this process.

The resistance results indicate that irreversible and reversible fouling (Rf) and gel resistance (Rg), represented 55% of the total resistance of DOW Filmtech NF90 membrane fouled with high DOC and hardness water. Surface morphology and depth of the gel layer was analyzed. The gel layer on the fouled membrane fouled at the water treatment plant represented 25% of the total thickness of the membrane. The laboratory fouled membrane, with higher flux resistance, had different surface roughness than that of the clean membrane. Therefore, the external surface of the gel layer can be related to this layer resistance.

Nisha Jha, M.Sc. candidate, Dr.Zsolt Kiss, post-doctoral fellow, and Ian Moran an undergraduate student at the University of Manitoba were conducting membrane laboratory testing and some membrane autopsy analyses. The author would like to thank National Science and Engineering Research Council (NSERC) of Canada for supporting this research.

References

1. Manitoba Office of Drinking Water http://www.gov.mb.ca/conservation/waterstewardship/odw/public-info/general-info/compliance_data/
2. AMTA, 2017. American Membrane Technology Associaton. www.amtaorg.com

3. M. Sadrnourmohamadi, C. Goss, B. Gorczyca, Water Sci. Technol. Water Supply **13**, 3 (2013)
4. M. Sadrnourmohamadi, B. Gorczyca, Water Res. **73** (2015)
5. B. Gorczyca, J. Ganczarczyk, Wat. Qual. Res. J. Canada **37** (2002)
6. M. Sadrnourmohamadi, B. Gorczyca, Sep. Sci. Technol. **50** (2015)
7. M. Hooshyar, M. Winnipeg, Eng. Report, University of Manitoba, Civil Engieering (2015)
8. K. Brezinski, M. Sadrnourmohamadi, B. Gorczyca, West. Canada Water (2015)
9. Y. Bessiere, B. Jefferson, E. Goslan, P. Bacchin, Desal. **249** (2009)
10. N. Geismar, P. Berube, B. Barbeau, Desalin. Water Treat. **43** (2012)
11. H. Wray, R. Andrews, P. Berube, Desal. **330**, (2013)
12. F. Chen, et al., Water Res. **48** (2014)
13. J. Tian, M. Ernst, F. Cui, M. Jekel, Water Res. **47** (2013)
14. P. Le-Clech, V. Chen, T.A. Fane, J. Membr. Sci. **284** (2006)
15. B. Gorczyca, I. Moran, *IWA Spec. Conf. June 22-24* (2016)
16. M. R. Wiesner, J.M. Laine, *Water Treatment Membrane Processes* (New York: McGraw-Hill, Inc. 1996)
17. D. Antelmi, B. Cabane, M. Meireles, P. Aimar, *Cake Collapse in Pressure Filtration* (Langmuir 2001)
18. J. Madeline, M. Meireles, R. Botet, B. Cabane, Wat. Sci. Technol. **53**, 7 (2006)
19. J. Madeline, et al. Restructuring of colloidal cakes during dewatering (Langmuir, 2007)
20. C. Parneix, J. Persello, R. Schweins, B. Cabane, *How do colloidal aggregates yield to compressive stress?* (Langmuir, 2008)
21. B. Josson, J. Persello, J. Li, B. Cabane, *Equation of state of colloidal dispersions* (Langmuir, 2011)
22. K. Kimura, T. Maeda, H. Yamamura, Y. Watanabe, J. Membr. Sci. **320** (2008)
23. M. Elimelech, S. Bhattacharjee, J. Membr. Sci. **145** (1998)
24. M. Cheryan, Ultrafiltration and Microfiltration Handbook (Technomic Publishing Co. Inc., Pennsylvania 1998)
25. F. Meng, S.R. Chae, A. Drews, M. Kraume, H.S. Shin, F. Yang, Wat. Res. **43** (2009)
26. J. Ganczarczyk, *IWA Spec. Conf. on Biofilm Systems* (1996).
27. A. Vahedi, B. Gorczyca, Wat. Res. **45**, 2 (2011)
28. A. Vahedi, B. Gorczyca, Wat. Res. **46**, 13 (2012)
29. A. Vahedi A. B. Gorczyca, Wat. Res. **53** (2014)
30. Pembina Valley Water Co-op. Annual Reports. http://pvwc.ca/ (accessed April 23, 2017)
31. N. Svenda, B. Gorczyca, New Orleans, American Water Works Association (2014)
32. N. Jha, B. Gorczyca, Z. Kiss, *An. Conference, WCWWA, Saskatoon, Sept.19-22* (2017)
33. B. Kaye, Winheim: VCH Verlagsgesellschaft (1989)

Contemporary aspects in the assessment of quality and treatment of swimming pool water

Mariusz Dudziak[1,*], *Joanna* Wyczarska-Kokot[1], and *Edyta* Łaskawiec[1]

[1] Silesian University of Technology, Faculty of Energy and Environmental Engineering, Konarskiego 18, 44-100 Gliwice, Poland

Abstract. Recent reports in the literature relate to the identification of various substances in the pool water, including low-molecular weight pharmaceuticals that can react with chlorine residues contributing to the formation of a large group of irritant and toxic compounds. These substances are not covered by standard monitoring. As part of this work, the authors present information on the methods for assessing the quality of the swimming pool water. Common processes used in swimming pool water treatment systems were described. Previous experience of the authors in the use of a detailed assessment of the quality of the swimming pool water on the basis of toxicity tests was presented. The authors have researched the application of pressure driven membrane filtration for the improvement of the efficiency of the swimming pool water treatment.

1 Introduction

The latest literature reports on the identification of various low-molecular substances, including the pharmaceutical ones, in swimming pool water [1]. These compounds may react with chlorine residuals, leading to the creation of a large group of irritative and toxic compounds.

Swimming pool water is commonly treated by coagulation, filtration and disinfection processes [1-3]. However, the problem related to the management of washings has not yet been solved. In the case of this waste flux there are no uniform procedures.

This paper presents information on the methods used for the assessment of the quality of swimming pool water and the types of processes that are used in swimming pool water treatment systems. The authors' past experience in using various toxicity tests for detailed swimming pool water quality assessments was demonstrated. The authors also researched the use of pressure membrane filtration (ultrafiltration) and determined if it can improve the effectiveness of a swimming pool water treatment process.

2 Swimming pool water quality assessment

2.1 Control

Legal and natural persons running businesses in the swimming pool industry are responsible for health and safety of the bathers. As a result of that the individuals that are in charge of swimming pool facilities should control the swimming pool water and ascertain that it is of the best quality.

Health and safety of the bathers is guaranteed by [1-3]: operational pool water treatment devices, maintenance of correct hygienic and sanitary conditions in the swimming pool facility, constant intake of clean water, health and hygiene of the bathers.

Microbiological and physicochemical contaminants present in swimming pool water may be harmful to health and cause sickness. The most common ailments affecting people who visit swimming pools with inadequate water treatment are [2-3]: conjunctivitis, ear inflammation, pharyngitis, gastro-intestinal illnesses, skin diseases (fungal infections, dermatitides, impetigoes, warts) as well as allergies and asthmatic respiratory symptoms.

The assessment of the swimming pool water quality performed by a person managing the swimming pool facility comprises [1-3]: visual assessment of the pool water quality in particular swimming pool basins, control of the operation of devices in the swimming pool water circuit treatment system by measuring the parameters of water quality, i.e. water pH, redox potential, free chlorine, water temperature and bounded chlorine.

The physicochemical requirements should only be an aid in the assessment of the swimming pool water. The most important are the results of microbiological tests and the evaluation of the state of the facility and devices after a sanitary inspection.

To correctly evaluate the quality of the swimming pool water and the effectiveness of its treatment system one should monitor not only the quality of water in the swimming pool basins, but also the quality of water from the mains (replenishing the losses and providing water for the retention tanks) and the quality of the filtrate which determines the filtration effectiveness.

* Corresponding author: mariusz.dudziak@polsl.pl

2.2 Microcontaminants

Among the organic microcontaminants present in swimming pool water, the literature concentrates primarily on the substances from the group of disinfection byproducts (DBP), that is chloroform and other trihalomethanes (THM) [2]. However, in 2002, Lambropoulou et al. [3] documented the presence of compounds coming from UV filters that are used in cosmetics (2-hydroxy-4-methoxyphenyl BP3) and (2-ethylhexyl-4-dimethylaminobenzoate OD-PABA). It means that these compounds were introduced to the swimming pool water by the bathers. Other researchers also confirmed the presence of compounds from UV filters in swimming pool water (Table 1) [4-8]. The analysis of Table 1 leads to a conclusion that the presence of UV filters in swimming pool water is a common occurrence and that the concentrations of particular compounds depend both on the type of the pool and its location (country). The concentration levels depend also on the type of the compound.

Ekowati et al. [8] tested the swimming pool water not only for the presence of compounds from UV filters, but also for a wide range of pharmaceutic substances. The water samples were collected from 17 swimming pools in sports centers and hotels in Catalonia, Spain. Among the pharmaceutical substances, 10 compounds were identified, i.e. atenolol, carbamazepine, hydrochlorothiazide, metronidazole, ofloxacin, sulfamethoxazole, paracetamol, ibuprofen, ketoprofen and phenazone. The highest concentration was found for a diuretic substance, hydrochlorothiazide, which amounted to 904 ng/dm^3. The most commonly detected substance was carbama-zepine which was found in more than half of all analysed water samples (53%, 27/51). The pharmaceutical substances were detected most frequently in spas.

On the other hand, Wenget al. [9] identified 3 substances of very different origin, i.e. N,N-diethyl-meta-toluamide - used as an insecticide, caffeine – natural psychoactive substance and tris(2-chloroethyl) - used as a flame retardant. It was also determined that the concentration of the insecticide was changing from season to season, reflecting the varying intensity of its use. It is very rare for this insecticide to be used in winter and, therefore, its highest concentration was detected in summer (approx. 2087 ng/dm^3). The concentrations of the other two substances, i.e. caffeine and tris(2-chloroethyl), were also the highest in summer. Caffeine may be introduced to swimming pool water with secretions, such as sweat and urine. The source of the flame retardant was not determined.

Microcontaminants present in swimming pool water may lead to its toxicity. Therefore, the assessment of swimming pool water should be aided by an ecotoxicological analysis.

2.3 Toxicological assessment

Ecotoxicological tools are commonly used in assessments of negative impact of chemical substances on live organisms. The indicator organisms used in biotests are characterized by varying sensitivity to a wide range of organic and inorganic compounds. The results of the biotests are obtained by observing the influence of toxicants on metabolism, decreased population, growth and development changes in the indicator organisms [10].

Among the test organisms there are bacteria, as well as animal and plant organisms. The analysis of the biotests results is based on toxicity classification of a particular sample. One of the most common toxicity classifications of samples was presented in Table 2.

Table 1. Concentration of UV filters in swimming pools.

Matrix	Concentration [ng/dm^3]			Country	Literature
	BP3	4MBC	OD-PABA		
Swimming pool	2400-3300	-	<900-2000	Greece	[3]
	4.2-5.7	5.4-6.9	-	Greece	[4]
	103-400	<146-330	<37	Slovenia	[5]
Baby pool	1200	10000	-	Germany	[6]
Swimming pool	-	600	-		
Public pool	<110	<200	<70	Spain	[7]
Private pool	<0.1-4.87	<0.4-5.60	<0.1		[8]
Indoor swimming pool	<0.1-15.17	<0.4-35.9	<0.1-2.0		
Outdoor swimming pool	<0.1-1.77	8.6-12.1	<0.1		
Indoor children	<0.1-1.97	8.1-45.4	<0.1		
Spa	<0.1-1.07	<0.4-69.3	<0.1		

Table 2. Classification of toxicity [10].

Effect, %	Toxicity class
<25	non toxic
25-50	low toxicity
50.1-75	toxic
75.1-100	high toxicity

The in-house research [11-13] on the use of biotests as a tool supporting the physicochemical analysis of the swimming pool water allowed to establish the preliminary procedures of their use.

The samples for the ecotoxicological tests may be taken from various places in the swimming pool water circuit what allows to control the water quality on every stage of its treatment, preliminary procedures of their use. The samples for the ecotoxicological tests may be taken from various places in the swimming pool water circuit what allows to control the water quality on every stage of its treatment, i.e. from the system water supply to waste fluxes from the washings of pressure filters. The samples may be collected to both glass bottles and plastic containers because it was proven that the material the containers are made of does not significantly influence the quality of the sample and the biotests results. The collected samples may be subjected to the biotests directly because there is no need for additional preparatory procedure. It is of vital importance that the biotests be accompanied by a basic physicochemical analysis of the sample. The tests results revealed that the increased toxicity was caused by free chlorine concentration in the tested samples [13]. As a result of that, it was necessary to establish a proper time gap or dechlorinate the sample, limiting the influence of chlorine on the biotest results.

During the in-house research [11-13] various indicator organisms at different levels of nature organization were used. Some of the tests involved the simultaneous use of a few indicator organisms that allowed to establish the differences in their sensitivity. For instance, while assessing the toxicity of samples [11], three indicator organisms were used, i.e. *Aliivibrio fischeri*, *Daphnia magna* and *Chaoborus flavicans* whose sensitivity to toxicants varied greatly. Fig. 1 shows the results of biotests for the sample of water from the mains, water from the swimming pool basin and waste water flux - washings. Microtox® test showed the toxic effect of water from the swimming pool basin and washings on *Aliivibrio fischeri* bacteria. Bioluminescence inhibition for these samples amounted to 65% and 53%, respectively. In the remaining biotests, the toxicity of the swimming pool water sample on the indicator organisms was low (*Daphnia m.* 8% and *Chaoborus f.* 22%). In the case of the washings sample, the tested organisms showed varying sensitivity to the ingredients present in it. The mortality rate of diptera larvae was approx. 90% what indicates that the sample is highly toxic (Table 2). However, *Daphia magna* test results show that the toxicity of the same sample is low (mortality rate of crustacea of approx. 42%). These results show that the sensitivity of indicator organisms varies greatly.

Among the indicator organisms employed in phytotests, aquatic plants from the *Lemnaceae family*, including *Lemna minor*, are also widely used. They are characterized by small size, simple structure and fast growth. These organisms are especially sensitive to surfactants of hydrophobic properties that concentrate at liquid-air interface [14]. During the in-house research [11], it was determined that phytotests allowed to specify which swimming pool water samples stimulate or inhibit the plants growth. This aspect is especially important as it allows to decide whether, for instance, the washings flux from the swimming pool installation may be used to water the greenery surrounding swimming pool facilities.

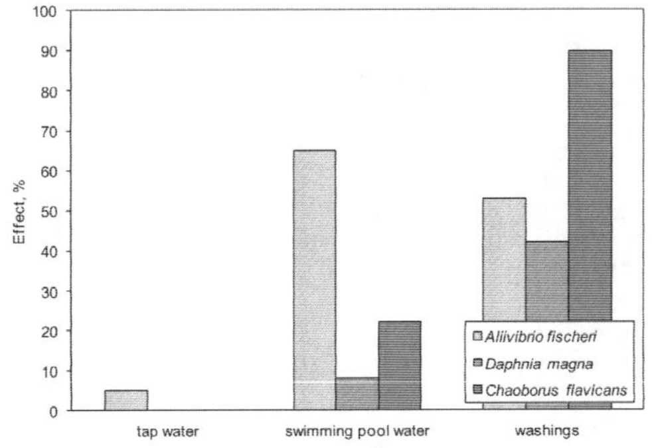

Fig. 1. Toxic effects in the Microtox® test, and in testing the survival of Daphnia magna and Chaoborus flavicans individuals [11].

Full toxicological evaluation may be an important tool supporting the standard physicochemical methods of swimming pool water quality assessment. The test results show that the sensitivity of indicator organisms varies greatly. The simultaneous use of a few biotests allows to establish more precisely the influence of chemical substances present in the tested samples on various live organisms. The phytotests, in turn, make it possible to determine both the stimulating and the inhibiting effect of the swimming pool water or washings samples on the growth and development of plants.

3 Processes and methods of swimming pool water treatment

The processes and methods for the treatment of water in the swimming pool circuit should allow for the accumulation of contaminants introduced to the swimming pool water by the bathers and provide a dose of disinfectant guaranteeing the removal of microorganisms from the swimming pool water basin.

The basic water treatment systems for sports and leisure swimming pools should consist of a filtration process, and specifically a surface coagulation run in filters with a sand and gravel bed, and a chlorine disinfection process, with the use of sodium hypochlorite (preferably, produced on site, in the process of membrane electrolysis).

The methods for the treatment of swimming pool water that allow to obtain water of quality and hygienic parameters compliant with the norms [15] include:

- filtration in sand filters + chlorine disinfection + pH correction,
- surface coagulation in sand filters, multilayered ones or with a layer of hydro-anthracite + chlorine disinfection + pH correction,

- surface coagulation + ozonization + sorption filtration + chlorination + pH correction,
- surface coagulation + ozonization + filtration in a multilayered bed (including the sorption layer) + chlorination + pH correction,
- ozonization + sorption filtration + chlorination + pH correction,
- surface coagulation in multilayered filters + ozonization of part of the circuit water flux + chlorination + pH correction,
- surface coagulation in multilayered beds + UV irradiation + chlorination + pH correction.

3.1 Filtration

Filtration is the main process in the system of swimming pool water treatment and its aim is to remove solids of varying dispersion from the water. In this way, the filtration protects the remaining parts of the installation against mechanical damage and clogging. It also influences the effects of the subsequent processes, disinfection and water pH correction.

Swimming pool water filtration systems usually utilize fast filters (pressure, multilayered, sand and gravel or with an additional layer of anthracite, active carbon, zeolite or ones filled with a glass bed) which filter water at 30 m/h [15].

Increasingly popular in the treatment of swimming pool water are vacuum filters (open or closed, with frame inserts and a filtration layer) that were previously used for industrial applications.

To make swimming pool water turbidity-free and compliant with current regulations [15], the use of pressure filters have to be combined with a coagulation process that supports water filtration (surface coagulation). In the case of open vacuum filters (mainly due to lower filtration speed, 4-11 m/h) there is no need for an additional support.

The open vacuum filters utilize diatomaceous earth, perlite or cellulose fibers that are washed over the filtration fabric and used as a filtration material. The filtering material in closed vacuum filters is quartz sand of varying granulation or, in justified circumstances, quartz sand and anthracite.

The in-house research [16] allowed to conclude that a swimming pool water treatment system could be accompanied by pressure membrane processes. The tests that have been conducted so far included an attempt to treat the washings by ultrafiltration process. The washings, being a wastewater flux, are characterized by high turbidity and high concentration of solutes, including disinfection byproducts and contaminants present in the chemicals used in the surface coagulation process. The research also described transport and separation properties of the membranes at various process parameters, i.e. type of the membrane material, membrane molar mass cut-off and transmembrane filtration pressure (ΔP). Fig. 2 shows the separating capabilities of membranes in terms of contaminants removal in a single ultrafiltration process described by absorbance retention coefficient in ultraviolet radiation UV_{254} and turbidity coefficient. During the ultrafiltration process, high values of turbidity retention coefficient were achieved (above 95%). Additionally, as the filtration time progressed, the effect of removing the compounds characterized by absorbance in ultraviolet radiation UV_{254} was increasing. By recovering the treated permeate flux, amounting to 50%, it was reduced by approx. 90%.

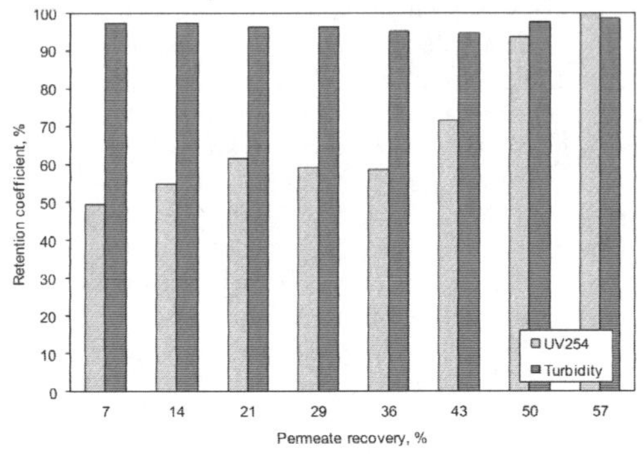

Fig. 2. Changes in the value of the retention factors of pollution in the unit process of ultrafiltration (ΔP = 0.2 MPa, membrane polimer - polyacrylonitryle) [16].

Membrane pressure processes allow to treat washings to a degree that they can be reused. However, the membrane transport conditions need to be improved by preliminary treatment processes.

3.2 Disinfection

Modern technologies for the treatment of swimming pool water, selection of appropriate chemical preparations, high filtering and disinfection efficiency guarantee that the pool water meets the strict requirements in this regard, providing water that is safe for health and comfortable to bathe in.

Although private and therapeutic swimming pools do not necessarily require the use of chemicals, it is not possible (for the time being) to avoid them in the case of public pools. The necessity to use chlorine is dictated by the need to prevent the contamination of water in the swimming pool built for "the masses".

Increasingly popular are systems aiding the final water disinfection with chlorine compounds and UV irradiation (medium and low pressure lamps) or ozonization. Introducing these solutions to swimming pool water treatment systems allows to reduce the amount of chlorine that is added at the final stage of the disinfection process [17]. This, in turn, allows to lower the saturation of chloro-nitrogen compounds that cause allergies, irritation of upper respiratory and digestive system, and which are also mutagenic and responsible for the characteristic unpleasant smell of swimming pool water [18-20].

The disinfectants used in the swimming pool technology include: sodium hypochlorite, chlorine gas, electrolytic chlorine, chlorine dioxide, as well as ozone and UV irradiation in combination with chlorine. Less commonly used are: bromine, iodine and silver ions in combination with chlorine. In swimming pools with mineral waters, such as rehabilitation and spa pools, increasingly popular are "low-salinity" technologies during whose operation NaCl that is added directly to swimming pool basin is subjected to electrolysis. A few swimming pools in Poland (so far, experimentally) support the process of swimming pool water treatment with devices emitting EMF and extended frequencies.

The disinfectant must not be chosen at random. The choice of disinfectant will depend on whether the treatment system is designed for public, leisure, aquapark, sports or a therapeutic swimming pool with, for instance, mineral-rich water. In the last case, it is important to thoroughly analyse the water and perform technological tests to ensure that the reaction of the chosen disinfectant with the compounds naturally present in the water is as mild as possible.

4 Conclusions

The identification of organic microcontaminants, including compounds from UV filters and pharmaceutical substances, in swimming pool water makes one consider the effectiveness of the modern technologies of water treatment. It is necessary to establish proper analytic procedures in this regard. The assessment of the quality of swimming pool water and waste fluxes from the treatment installation should by aided by an ecotoxicological analysis. The sets of toxicological tests should also be researched before the selection as these will allow to assess the real effects of the contaminants together with their antagonistic and synergic influence on properly selected test organisms what will, in turn, allow to interpolate the results to a human organism. While searching for a new technology for the treatment of swimming pool water, it is suggested to consider membrane pressure filtration, including the ultrafiltration process.

References

1. M.G.A. Keuten, F.M. Schets, J.F. Schijven, J.Q. Verberk, J.C. van Dijk, Water Res. **46,** 3682 (2012)
2. S.J. Judd, S.H. Black, Water Res. **34**, 1611 (2000)
3. D.A. Lambropoulou, D.L. Giokas, V.A. Sakkas, T.A. Albanis, M.I. Karayannis, J. Chrom. A **967**, 243 (2002)
4. P. Giokas, M.S. Díaz-Cruz, D. Barceló, Sci. Total Environ. **519**, 518 (2015)
5. P. Cuderman, E. Heath, Anal Bioanal Chem. **387**, 1343 (2007)
6. C. Zwiener, S.D. Richardson, D.M. de Marini, T. Grummt, T. Glauner, F.H. Frimmel, Environ Sci Technol. **41**, 363 (2007)
7. L. Vidal, A. Chisvert, A., Canals, A. Salvador, Talanta **81**, 549 (2010)
8. Y. Ekowati et al. Environ Sci Pollut R **23**, 14431 (2016)
9. S. Weng, P. Sun, W. Ben, C.H. Huang, L.T., Lee, E.R. Blatchley, Environ. Sci. Technol. Lett. **1**, 495 (2014)
10. S. Werle, M. Dudziak, Przem. Chem. **92**, 1350 (2013)
11. E. Łaskawiec, E. Kudlek, M. Dudziak, J. Wyczarska-Kokot, Gas Water San. Tech. **9**, 330 (2016)
12. E. Łaskawiec, M. Dudziak, J. Wyczarska-Kokot, Ecol. Eng. **50**, 210 (2016)
13. E. Łaskawiec, J. Wyczarska-Kokot, M. Dudziak, JCEEA **33**, 143 (2016)
14. B.S. Mohan, B.B. Hosett, Environ. Pollut. **98**, 233 (1997)
15. DIN 19605: 2016-05: Festbettfilter zur Wasseraufbereitung – Aufbau und Bestandteile
16. E. Łaskawiec, M. Dudziak, J. Wyczarska-Kokot, COBRABiD **4**, 238 (2016)
17. J. Wyczarska-Kokot, RI **12**, 96 (2009)
18. J. Wyczarska-Kokot, Ochr. Sr. **36**, 37 (2014)
19. J. Wyczarska-Kokot, Ecol. Chem. Eng. S **22**, 27 (2015)
20. J. Wyczarska-Kokot, Ecol. Chem. Eng. S **23**, 447 (2016)

Ozone as one of the factors affecting the safety of water supply

Józef Natonek[1,*]

[1]Jaworzno Waterworks Ltd, 43-600 Jaworzno, 34 Św. Wojciecha St, Poland

Abstract. Ozone, due to its strong oxidizing properties, is often used in high efficiency water treatment processes. It is not only used for water disinfection, but also for carboxylic acid, a product of its reaction, which highly contributes towards increasing biodegradability of organic substances in water, thus leading to higher quality of drinking water. This paper presents results of using mobile ozone devices as a way of local microorganisms elimination and sediments removal from WSS.

1 Introduction

Most waters intended for drinking by man that are taken in, both underground and surface waters, require more or less complex water treatment processes. One of such processes is oxidation, which is performed for many purposes, for instance [1-5]:

- removal of iron, manganese and nitrogen compounds from water,
- removal of hydrogen sulphide and sulphides from water,
- oxidation of compounds causing water to have colour, taste and odour,
- oxidation of micro-contaminations and natural organic matter,
- neutralisation of phytoplankton and zooplankton,
- facilitation of coagulation as a result of colloid destabilisation,
- disinfection of water.

Ozone, the reduction potential of which is 2.07 V, plays a significant role in those processes. The key characteristic of ozone is its eco-friendly effect eliminating the formation of adverse oxidation by-products generated in processes with use of chlorine compounds (chlorinated by-products). What is more, ozone disintegrates, coming back to pure oxygen, during ozonisation.

Due to its properties, ozone is commonly used in the treatment of water indented both for municipal and industrial purposes. In Poland, in 1959 – nearly 66 years after ozone was used in Dutch waterworks for the first time, ozone-based water disinfection was implemented in the waterworks of Kraków, in the Bielany water treatment plant. In practice, ozonisation is performed as the preliminary or intermediate oxidation process when using water supply systems. In those processes, ozonisation – being an advanced oxidation process – not only catalyses chemical reactions occurring in water, but it also – as a result of colloid depolarisation – supports the agglomeration of particles, and so increases the processing effectiveness of coagulation and filtration and decreases the dose of used coagulants by even 50%. Moreover, in order to improve the safety of supplying water to consumers, ozone can be used to remove other contaminations such as phenols, detergents, pesticides, coordination complexes, pharmaceuticals or hormones. Apart from this, ozone is frequently the only and the most effective solution in processes of improving organoleptic characteristics of water, i.e. odour, taste or colour. Using ozone in water treatment contributes to reduction of applied chemical disinfectants considerably. Numerous studies [6,7] show that in treatment processes, ozone allows obtaining biologically stable water, which is one of the requirements for the possibility to depart from chemical disinfection of water. All that we can achieve using ozone brings us closer to an eco-friendly water treatment system, where all chemical reagents are eliminated or minimised.

A significant use-related problem of water supply systems is widespread secondary water contamination when water is transported to the consumer [8-11,12]. The biofilm forming in the water pipes network is a considerable habitat of microorganisms, including pathogenic ones. In those circumstances, changing hydraulic conditions of use of water pipe network increase the risk of microbiological water contamination. This can be exemplified by a range of cases where biologically contaminated water, with *coli* group bacteria, *Escherichia coli* bacteria or *Enterococcus faecalis* bacteria, was supplied to consumers [11, 13-19]. The issue of microbiological contamination in water is often associated with the necessity to stop supplying water to the consumer or inform them about the fact that water is unfit for drinking. In such situations, water authorities must provide alternative water supply sources, identify the cause of contamination and take remedial action, such as flushing and disinfecting the water pipes network.

* Corresponding author: nato10@poczta.onet.pl

Strong oxidation properties of ozone in a wide range of water reaction, short reaction time, no by-products detrimental to man and the environment, strong destructive properties in relation to pathogenic organisms and spores are qualities that allow this agent to be used in water supply system use operations. A significant part of this method is first and foremost short time required for effective disinfection of waterworks, which can translate into a considerably shorter time of water supply stoppage after its secondary biological contamination.

2 Study subject

Use of water pipes network as vast as that of Jaworzno is quite a challenge. In the city with the area of nearly 152 km^2 and waterworks of total length of ca. 600 km, a lot of conditions and factors decide whether water of proper quality is being supplied. Despite high quality of water pumped in waterworks, the aging, material-related and hydraulic structures have an effect on maintenance of physicochemical and microbiological stability of potable water when it is being supplied to the consumer. The Jaworzno waterworks is built of PE, PVC, cast iron and steel. The material diversity facilitates origination of many threats of deterioration of quality of water supplied to the consumer, for instance due to biofilm formation.

To increase the guarantee of high quality water supply and to ensure that water of stable parameters is supplied, MPWiK Sp. z o.o. in Jaworzno takes a range of actions aiming to improve water quality safety in the very waterworks. To that end, threats related to the following aspects are eliminated:
secondary water contamination in waterworks due to pipeline failure,

- proper disinfection of newly build waterworks sections before they are put into use,
- periodical disinfection of waterworks terminals with the longest water retention time.

If distributed water deteriorates in quality, particularly in terms of microbiology, a significant factor having an effect on the safety of water supplies in the system is the time needed to take and complete emergency action. The disinfection agents used so far (chlorine compounds) in the company have been effective, but they act for a long time (at least 1.5 day) and can be burdensome for the consumer. Therefore, in line with the policy of constantly improvement in quality of water supplied to Jaworzno inhabitants and in accordance with risk management procedures, an innovative and effective manner for water disinfection in waterworks lines, i.e. field supply system ozonisation, was implemented in the second half of 2016 in Jaworzno water supply system. Disinfection of water in the system with ozone considerably reduces the time of procedure (to several hours), and so increases consumer health safety and allows the waterworks infrastructure operator to remove chemical disinfection effects in the field as in the case of disinfection with traditional sodium hypochlorite.

3 Principles of use of mobile waterworks flushing and disinfection system

Mobile water ozonisation system "SPID" operates automatically. The nominal efficiency of the system is 1.5 m^3/h. The SPID ozonisation system is an assembly of technological unit elements providing the production of ozone (module for production of ozone from air in plasma blocks) in places of its application to the waterworks as well as an assembly of devices of a module for introducing ozone in water. Component parts of the modular assemblies of the mobile flushing and disinfection system SPID providing its effective operation in the field are as follows (Fig. 1):
oxygen production module – oxygen producer TOPAZ PLUS or ULTRA,

- ozone production ozone – one-module ozone generator, e.g. 1Z-WO60F-GT,
- water preparation module – two contact columns no. 1 and 2,
- module for introducing ozone in water: circulating pump PA, injector, ozone separator, quick stop valve ZE1, static mixer,
- highly ozonised water pumping module,
- module for measurement of residual ozone in water – on-line measuring device KRYPTON NEON DES with feedback control,
- module for measurement of ozone in the air – on-line measuring device K 100 W GAS,
- destruction module with redundant ozone blow-through.

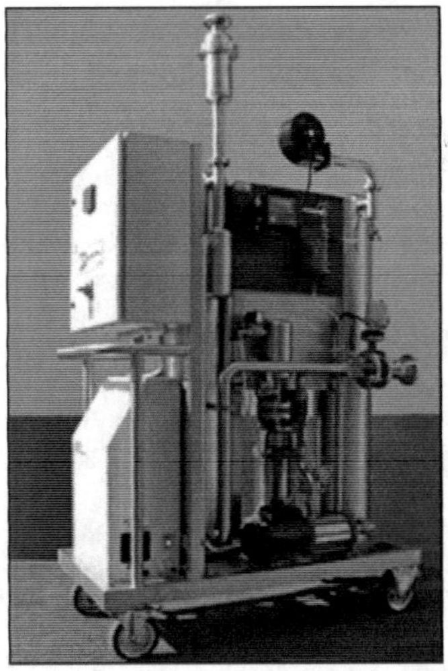

Fig. 1. LSPID - mobile system for flushing and disinfecting of water pipe network.

System of automatic control over operation of the mobile SPID device according to a strict configuration of individual unit processes allows a considerable reduction in the number of members of the operation-and-renovation team for operation of water supply system. After contact columns are filled with water, the valve allowing water to the ozonisation system closes automatically. Before the contact columns, a system for introducing ozone with the circulating pumps. Ozone is produced in the ozone generator out of the oxygen produced in the oxygen producer TOPAZ PLUS. The generated ozone is supplied by means of pressure to the ozone separator, equipped with safety measures protecting the ozone separator from flooding. The system for protecting the ozone generator from flooding is a three-degree protection system:

- Degree I – check valve in the injector,
- Degree II – check valve directly before the separator,
- Degree III – separator equipped with a conductometric probe cooperating with quick electric valve ZEO1 cutting off ozone supply.

The ozone from the separator is sucked under pressure by the injector as a result of water flowing through it. It is there that ozone is mixed with water from the second contact column, directed to individual component parts of the module for introducing ozone to water with circulating pump, for the first, preliminary time. The water, preliminarily ozonised in the injector, is pumped into the conduit feeding contact column no. 1, where it is thoroughly mixed with raw water in the static mixer. In the contact columns, water is saturated with oxygen. In column no. 2, water is degassed as a result of desorption of ozone, which is discharged outside with a fan by the ozone destructor. After contact column no. 2, the water is pumped into a selected part of the water supply system to disinfect or flush it. The ozone content in water supplied to the waterworks for disinfection fluctuates, on average, from 0.5 mgO_3/L to 1.0 mgO_3/L or more, depending on the required effects in terms of maintenance and operation of the waterworks.

In order to obtain high effects of waterworks disinfection, each and every time the process duration is to be selected well, which is determined both by the diameter and length of the conduit. The dose of ozone is also determined by the residual ozone content in the stream of water flowing out of the system being disinfected at the optimal level of 0.2-0.3 mgO_3/L.

4 Study methods and discussion of results

The effectiveness of the operation of the mobile SPID system in the water supply system of Jaworzno was assessed in the modular system in June 2016. The water was examined in accordance with the recommended analytical procedures:

- temperature – acc. to PB-54 edition 01/22.04.2009,
- general number of microorganisms at 22±2°C after 72 hours – acc. to PN-EN ISO 6222:2004,
- *Coli* group bacteria – acc. to PN-EN ISO 9308-2:2014-06,
- *Escherichia coli* – acc. to PN-EN ISO 9308-2:2014-06,
- *Enterococci* – acc. to PN-EN ISO 7899-2:2004,
- *Clostridium perfringens* (along with spores) – acc. to PB-28 edition 02/11.01.2010,
- *Pseudomonas aeruginosa* – acc. to PB-29 Edition 01/20.03.2013,
- ozone – acc. to PB-27 Edition 01/17.01.2013,
- turbidity – acc. to PN-EN ISO 7027:2003, Point 6,
- iron – acc. to PB-05 Edition 01/25.03.2010,
- manganese – acc. to PB-06 Edition 01/25.03.2010,
- bromates – acc. to: PN-EN ISO 15061:2003.

The modular system was built of PE pipeline with the diameter of 63 mm and length of 150 m. The system was filled with water from the local pond, which water was held in the system for two days in a sun-exposed place in order to obtain highly excessive microbiological quality in water (Table 1). After that time, the system was flushed and pipeline disinfection began by setting the starting ozone dose in the SPID device at ca. 0.3 mgO_3/l. The disinfection effect was monitored for 4 hours, when microbiological parameters and residual ozone content were being measured in water samples. The studies showed that over the first 20 minutes from pumping the ozonised water in the contaminated system, considerable reduction in microbiological parameters was achieved, i.e. the overall number of organisms dropped by 75% at 22±2°C. After 72 hours, the overall number of organisms went down by 73% at 36±2°C and after 48 hours there was a 85% and 87% removal of *Coli* group bacteria and *Escherichia coli* bacteria, respectively.

Table 1. The microbiological parameters of water quality in the model system – research time 03.06.2016

Sample time	Temp.	Number of micro-organisms at 22°C after 72 hours	Number of micro-organisms at 36°C after 48 hours	Coli group bacteria	Escherichia coli	Enterococci	Clostridium perfringens	Pseudomonas aeruginosa	Ozone
	°C	CFU/mL	CFU/mL	CFU/100 mL	CFU/100 mL	CFU/100 mL	CFU/100 mL	CFUL/100ml/L	mg O_3/L
09:55	16	>300	>300	88,5	88,5	0	0	0	Raw water
10:14	16	74	82	13,7	11,1	0	0	0	0.21
10:35	16	4	2	0	0	0	0	0	0.33
10:50	16	4	4	0	0	0	1	0	0.12
11:05	16	0	6	0	0	0	0	0	0.1
11:35	16	28	19	0	0	0	3	0	<0.02
11:40	16	16	0	0	0	0	0	0	<0.02
14:00	16	13	3	0	0	0	0	0	<0.02

After 40 minutes from the beginning to the end of the experiment, a stable effect of the process was achieved, characterised mostly by microbiological parameters of water quality at 0 CFU/100 mL (Table 1). In obtained water samples only, the overall number of macro-organisms both at 22°C after 72 hours, and 36°C after 48 hours, was from several to over a dozen colony forming units.

5 Conclusions

Preliminary studies of the technological effect of mobile SPID device to flush and disinfect waterworks showed significant limitation of threats for the consumer when microbiological contamination occurs in the supplied water. The effect was achieved thanks to reducing the duration of disinfection and flushing by many times: compared to commonly used chlorine compounds, from several days (2-1.5 days) to several hours (2-4 h). Moreover, the studies showed that:

- the initial ozone content at ca. 2 mg O_3/L with the flow of water at 2-3 m^3/h allows obtaining the first disinfection effect already after 20 minutes in the operation of the system with no side effects of disinfection (no bromate formation),
- the method is fully accepted by the consumer (residual ozone in water is at 0.02 mgO_3/L) and supervisory bodies such as Poviat Sanitary Inspectorates,
- ozone disintegrates to pure oxygen in 100% after the processes and the devices of technological infrastructure do not require another flushing.

References

1. A. Kowal, M. Świderska-Bróż, *Oczyszczanie wody - podstawy teoretyczne i technologiczne, procesy i urządzenia (*PWN, Warszawa 2007)
2. J. Nawrocki, Ochr. Sr. **27**, (2005)
3. R. Broseus, et al., Water Res. **43**, 18 (2009)
4. M.A. Oturan, J.J. Aaron, J. Critical Reviews Environ. Sci. Technol. **44,** 23 (2014),
5. M. Pawlita-Posmyk, M. Wzorek, Chemik, **70**, 10, (2016)
6. D. van der Kooij, et al. *Distributing drinking water with a low of zero disinfectant residua, operational and biological aspect* (UKWIR, London (2003)
7. P. Smeets, et al., Drinking Water Eng. Sci. **2**, (2009)
8. I Zimoch, E. Łobos, Desalin. Water Treat. **52**, (2014)
9. B. Kołwzan, Ochr. Sr. **33**, 4 (2011)
10. J. Wingender, Int. J. Hyg. Environ. Health, **214**, 6 (2011)
11. I. Zimoch, Ochr. Sr. **29**, 4 (2007)
12. *Bakterie w wodzie* (Accessed on 12.12.2014).
13. *http://czestochowa.naszemiasto.pl/artykul/bakterie-grupy-coli-w-wodociagu-koniecpol-nie-pijcie-wody* (Accessed on 19.11.2016)
14. http://www.dziennikpolski24.pl/artykul/3682072,bakterie-w-wodzie,id,t.html(Accessed on 12.12.2014)
15. I. Zimoch, Ochr. Sr. **29**, 3 (2007)
1. J. Boguniewicz-Zablocka, I. Klosok-Bazan, V. Naddeo, Environ. Sci. Pollut. Res. (2017) (doi: 10.1007/s11356-017-0608-8)
16. http://pssechrzanow.wsse.krakow.pl (Accessed on 03.09.2016)
17. https://gazetalubuska.pl (Accessed on 03.09.2016r.)
18. M. Bożym, I. Kłosok-Bazan, M. Wzorek, Pol. J. Environ. Stud. **27**, 3 (2018)
19. M. Gono, M. Kyncl, R. Gono, I. Kłosok-Bazan, PE **11**, 89 (2013)

Hardness of water analysis at the Grotowice intake situated near Opole

Iwona Klosok-Bazan[1,*], *Mateusz* Filipowski[2], and *Michał* Stojak[2]

[1]Opole University of Technology, Mechanical Department, 45-271 Opole, Mikolajczyka 5 St., Poland
[2]Wodociągi i Kanalizacja Sp. z o.o. w Opolu, Opole, 45-222 Oleska 64 St., Poland

Abstract. The determination of raw water hardness subsequently treated and used for domestic water supplies is a useful parameter to characterize the present total dissolved solids as well as for calculation dosages for lime-soda softening application. Because water hardness concentrations have not been proven to be health related, the final achieved level is mainly an economic function. To identify existing trends, the changes in water hardness in Grotowice intake for the years 1999-2017 have been analysed. The obtained results indicate clearly that the water hardness in three wells shows a slight downward trend, but on the other hand, one trend is rising in another direction.

1 Introduction

Hardness is one of the most noticeable indicators of water quality in the tap. This feature is treated by many customers as a treat to health due to more and more repeated but not proved thesis that hard water results is creation of renal calculi. Meanwhile, as the results of research conducted by the World Health Organisation show [1], both elements which cause the hardness i.e. calcium and magnesium, have a positive impact on human health and their deficit in drinking water may result in electrolyte imbalance. Water hardness is not regulated in the EU law. Under the Polish regulations, hardness of drinking water should be between 60-500 $CaCO_3/dm^3$, while water which is fully in compliance with the Regulation of the Minister of Health on quality of water intended for drinking by people may be not accepted by customers [2] due to excessive precipitation of sediments during its heating process and cooking. Depending on the spring quality, water which used by 4 members family can cause precipitation of even 70 kg of scale during the year. The amount of precipitated scale is not always proportional to the calcium and magnesium concentration. Precipitation of scale and sediments from water is connected with the so-called temporary hardness i.e. hardness caused by the presence of dissolved salts, mainly hydrogencarbonates which are precipitated during the heating. High carbonate hardness results in many operating problems in the waterworks system as well as in terms of recipients [3-15].

Decrease of calcium carbonate precipitation tendency as a result of water heating is technologically subject to decrease of alkalinity. It is the declination of alkalinity, i.e. decrease of HCO_3 anions which influences the decline of the amount of sediments precipitated from the water, however it is not without influence on metal element of waterworks system and internal installations. Therefore, while considering the possibility of introducing water softening systems, it is important to be aware of the negative effects of such actions, manifested in an increase in water aggressiveness level [2,7]. Meanwhile, the water in the tap should be stable. Langelier Index (I_L) and Ryznera Index (I_R) are often used to evaluate water stability. Unfortunately, these indices allow mostly for comparison of corrosivity of various waters [7,12] and, first and foremost, rough estimation of aggressive carbon dioxide concentration. Meanwhile, water stability is determined on the basis of aggressive carbon dioxide (stable water cannot contain aggressive carbon dioxide CO_2). Stability criterion is determined by water reaction in the state of pH_S saturation. Stable water is water which neither dissolves nor precipitates calcium carbonate. Water is stable when I_L e is equal to 0, which suggests that water is not aggressive and has no tendency to create protective $CaCO_3$ layers. When I_L is below 0, water is unsaturated and has aggressive properties but when I_L is higher than 0, water has a tendency to form $CaC0_3$ sediments but it is not aggressive [16]. Taking the abovementioned into consideration, making proper decision regarding whether and with what methods the water extracted from the intake must be softened is a substantially complex problem. Analysing the available literature [17-20] on the use of water softeners, the precipitation methods using various types of reactors are most commonly used in drinking water supply systems. Application of such a technology requires incurring significant investment expenditures and production of water will require high operating expenses. In consequence, it will induce an increase of price of water delivered to the recipients. However, when the hardness level is "dangerously" close to the level of 500 mg $CaCO_3/dm^3$, the investment

* Corresponding author: i.klosok-bazan@po.opole.pl

becomes inevitable. The authors of this work, on the basis of available archival data, decided to analyse changes of the water hardness level, the intake of which is in Grotowice near Opole, to establish the existing tendencies. A significant increase tendency will mean the necessity of inclusion of a new technological system in the investment plans which will serve as a water softener.

2 Description of research area

Grotowice water intake uses underground water from the Main Underground Water Reservoir GZWP 333. The area of intake location is a fragment of the Opole Plain. Carboniferous, Permian, Triassic, Cretaceous, Tertiary and Quaternary rocks are part of the geological structure of the described area. The underground reservoir is created by formations of inferior and medial buntsandstein, röt formation and muschelkalk. Currently, the intake includes 4 wells marked as 1A, 1B, 3B and 6B. Wells 1B and 6B have been functioning since 1995, and 3B and 1 A since 2008 (Fig. 1).

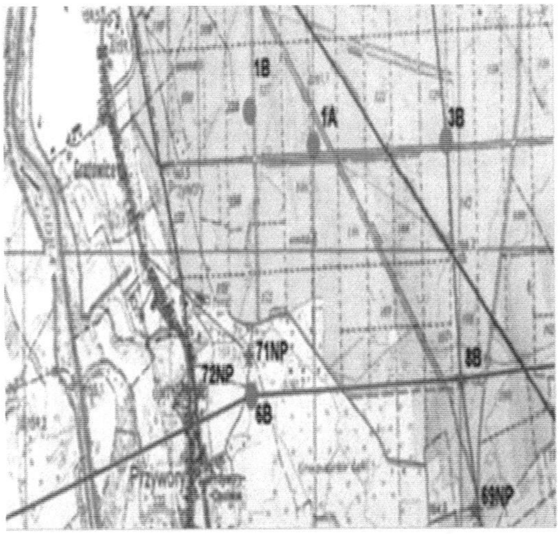

Fig. 1. Wells location in Grotowice intake.

Wells 1b and 6b have been used continuously and well 3B is included in case of higher demands. Well 1A is an emergency well. Raw water is a mixture of water from particular wells with a significant share of water from wells 1 B and 6B. Raw water pumped from Grotowice intake is delivered to the water supply network without treatment, however for the purpose of bacteriological protection, the water is disinfected using sodium hypochlorite. This water has very good quality, neutral reaction of ca. 7.5 and alkalinity ranging from 4.2 to 5.4 mval/dm^3. Water is characterised by low concentration of iron, it does not need application of deironing processes. Unfortunately, water analyses indicate presence of nitrogen compounds the concentration of which, in the analysed period, reached values in the range of: 0.01-0.061 mgNH_4^+/dm^3; 2 –23 mgNO_3/dm^3 and 0-0,024 mgNO_2/dm^3. The characteristic feature of Grotowice water is its hardness reaching 440 mg $CaCO_3$/dm^3. Relatively high hardness connected with the lack of thermal stability makes this parameter cause the highest number of customers' complaints. The water has tendency to precipitate sediments when heated [17]. The fact that the source is subject to strong anthropogenic impact caused by the active calcium excavations the exploitation of which has a significant impact on the changes of the level of water surface in the reservoir, is important from the point of view of the analysis of water hardness changes at the Grotowice intake. Both the quality of raw water as well as the quality of water in the particular wells is systematically monitored by Wodociągi and Kanalizacje Sp z.o.o in Opole.

3 Materials and methods

The quality of raw water in Grotowice intake is affected by the water parameters of the particular intake wells. In order to determine the tendency of changes, water quality in all exploited wells of Grotowice was analysed. Available archival data regarding analyses of water hardness were collected. The first stage of the analysis included determination of the characteristic concentrations for the analysed parameter: minimum, maximum and average values as well as standard variations. Potential seasonal changes of water hardness levels were analysed at the next stage of the paper by means of analysis of average concentration in the particular months in each of the analysed periods. In order to calculate the average hardness levels, the harmonic mean calculations were applied as, according to the authors of this article, it fits this type of analysis well due to the fact that the analysed data are relative values with identical signs. Then, by means of application of harmonic mean, average annual level of water hardness in the particular wells was calculated and the obtained data were used to specify the trend function. Graphical analysis was applied, then it was examined whether the implementation of dependent variable does not negate the dynamic properties of the function adopted initially as a result of the analysis. The available results of water quality analyses conducted by Wodociągi i Kanalizacje Sp. z o.o. in Opole were used as source data. Data obtained in 1997-2017 were interpreted for well 1B and 6B, and data from 2009 were interpreted for well 3B and 1A.

4 Results and discussion

Water hardness in the particular wells of Grotowice intake is subject to fluctuations. Those fluctuations are of temporary nature and do not show any significant identifiable tendencies, they appear unexpectedly and the reasons for which they occur are difficult to identify. Surely, it can be said that they are not connected with the changes of water volume collected from the analysed wells (Fig. 2-Fig. 4).

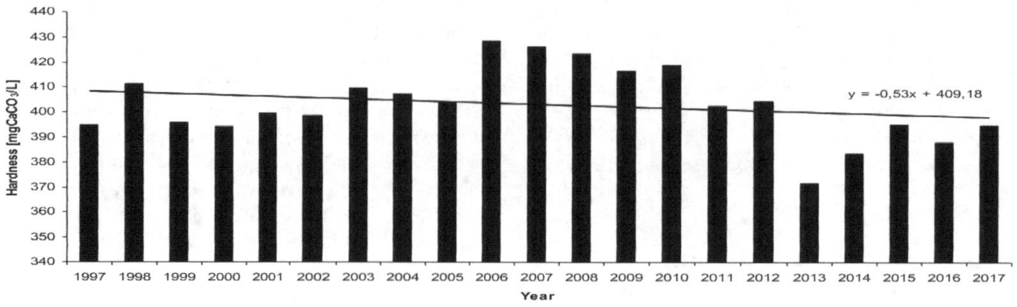

Fig. 2. Average hardness in years1997 -2017 in 1 B Well, with trend function.

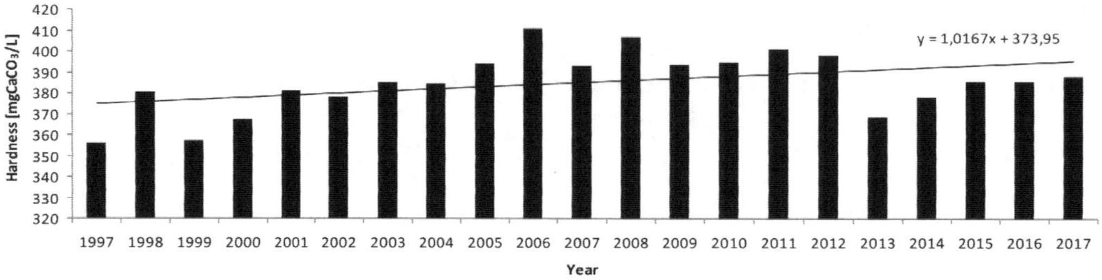

Fig. 3. Average hardness in years 1997 - 2017 in 6 B Well, with trend function.

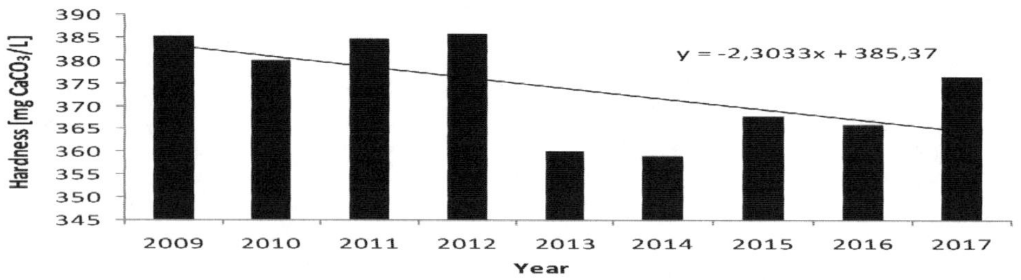

Fig. 4. Average hardness in years 2009 - 2017 in 3 B Well, with trend function.

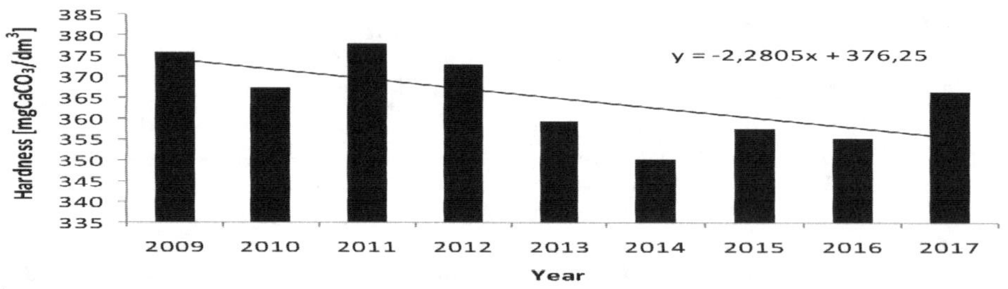

Fig. 5. Average hardness in years 2009 - 2017 in 1 A Well, with trend function.

The highest level of hardness for well 1 B was recorded in 2008, and the highest average hardness throughout the year was detected in 2006.

An analysis of hardness level in the particular months of the period from 1997 to 2017 did not reveal any seasonal change tendencies. The determined trend line for the average annual values shows a slight decline (Fig. 2). The well in which the highest concentration of the analysed parameter was found is well 6B. Maximum hardness level of 535 mg $CaCO_3/dm^3$ was recorded in June of 2008, however, analyses preceding this high level were respectively 402 and 460 $CaCO_3/dm^3$ with a monthly average of 420 $CaCO_3/dm^3$. In case of well 6B, the trend line for the average annual hardness value is, in its nature, an increasing function. This is the only well in which the calculated trend function is increasing (Fig. 3). Data regarding average annual level of water hardness, together with the marked trend line are presented in (Fig. 4). Similarly to the case of well 1B, the determined trade line takes the form of a decreasing function. The highest fluctuations of the monthly average value of the analysed parameter were detected in well 1A; they are between 332-398 mg $CaCO_3/dm^3$. This is probably due to the fact that the wells are included in operation in a cyclic manner. With the current level of production, this well is launched mainly during water collection for monitoring purposes. The line for wells 1B and 3B, determined for the average annual hardness levels in well 1A, is decreasing in nature.

6 Conclusions

The statistical analysis based on interpretation of actual archival data has shown that the water hardness level of three wells of Grotowice intake (1B, 3B and 1A) show a slightly decreasing tendency, but water hardness in well 6B shows an increasing tendency. Investigating the results of the particular analyses performed between 1997-2017, one should note the high variability of values obtained in the particular determined points. Taking into consideration the fact that the laboratory where the analyses were performed holds an accreditation, the argument of significant measurement errors was rejected. Therefore, the water in the source shows significant variability of the water hardness parameter and it must be emphasised that those fluctuations occur with short intervals. The thesis connected with the influence of the method of mining of the calcium minerals on the variability of collected water was not confirmed and requires detailed analyses.

References

1. Hardness in drinking-water. Background document for development of WHO Guidelines for Drinking-water Quality. World Health Organization, 2011
2. W. Sawiniak, I. Klosok-Bazan, J. Szczepankiewic, Instal **12**, (2014)
3. M. Mons, J. Van Dijk, D. Gatel, Global Water Res. Coalition, London, (2007)
4. J. Nawrocki, S. Biłozor, *Uzdatnianie wody Procesy chemiczne i biologiczne* (PWN Poznań 2000)
5. I. Klosok-Bazan, Ochr. Sr. **36,** 2 (2014)
6. I. Zimoch, Ochr. Sr. **30**, 3 (2008)
7. K. Kuś, F. Piechurski, D. Okoń,. Ochr. Sr. **51**, 4 (1993)
8. Ł. Weber, Techn. wody, 21 (2013)
9. I. Zimoch, Ochr. Sr. **34**, 4 (2012)
10. B. Godskesen, M. Hauschild, et al., J. Envir. Manag. **105** (2012)
11. I. Zimoch, E. Szymura, K. Moraczewska-Majkut. Desalin. Water Treat. **57**, 3 (2016)
12. J. Hoek, B. Dijkman, et al., Water Sci. Technol. **39**, 5 (1999)
13. I. Zimoch, Ochr. Sr. **29**, 4 (2007)
14. M. Bożym, I. Kłosok-Bazan, M. Wzorek, Pol. J. Environ. Stud. **27**, 3 (2018)
15. M. Pawlita-Posmyk, M. Wzorek, Chemik, **70**, 10, (2016)
16. J. Hofman, O. Kramer, P. Van Der Hoek, Water **21**, (2007)
17. M. Mons, J. Van Dijk, Drinking water hardness: Reasons and Criteria for Softening and Conditioning of Drinking Water, Global Water Research Coalition, London, (2007)
18. B. Van der Bruggen, H. Goossens, P. Everard, P. Stemge´e, W. Rogge, J. Environ. Manage. **91**, (2009)
19. J. Boguniewicz-Zablocka, I. Klosok-Bazan, E. Podgorni, Forum on Current issues in water treatment and water distribution, Gliwice, (2016)
20. I. Zimoch, E. Szymura, K. Moraczewska-Majkut. Przem. Chem. **94**, 2 (2015)

Approach to tap water safety analysis in terms of biostability

Barbara Tchórzewska-Cieślak[1], *Dorota* Papciak[2], *Katarzyna* Pietrucha-Urbanik[1,*], and Andżelika Pietrzyk[2]

[1]Rzeszow University of Technology, Faculty of Civil, Environmental Engineering and Architecture, Department of Water Supply and Sewerage Systems, Al. Powstancow Warszawy 6, 35-959 Rzeszow, Poland
[2]University of Technology, Faculty of Civil, Environmental Engineering and Architecture; Department of Water Purification and Protection, Al. Powstancow Warszawy 6, 35-959 Rzeszow, Poland

Abstract. The subject of the work is the analysis and assessment of the risk of biological instability of water. The lack of water stability causes the increased susceptibility of the distribution system to secondary microbial contamination of water and constitutes a hazard for consumers' health. The risk is expressed as the loss of water supply safety and distinguishes a failure of not meeting certain water quality parameters that can influence physico-chemical parameters and the bacteriological quality of the water supplied to the consumers. In the paper the method of analysing and evaluating the risk of loss of biostability of tap water is presented. The presented analysis was performed on the basis of the operating data from the water treatment plant.

1 Introduction

The primary and basic subject to which the concept of water safety concerns is the consumer. A measure of safety in relation to water consumers is risk, defined as the probability of undesirable event occurrence, whose consequence may be a real threat to the lives or health of water consumers [1-4].

When analysing the functioning of the water distribution subsystem (WDS) in terms of the protection of water consumers mainly the resistance of system to various types of undesirable events, that may occur in the subsystem is taken into account [5,6]. The consequence of these events may be break in water supply or the so called secondary water pollution in the water supply network. Hydraulic conditions in the WDS influence changes in the physicochemical and bacteriological water composition. The important problem that arises in many urban water supply systems is also the significant pipes oversizing which results in a decrease in the flow velocity of water and, consequently, in unfavourable flow conditions that cause deterioration of water quality in the water supply network.

The main objective of the paper is to present a methodology for analysing and assessing the risk of biological instability in terms of the water consumers safety.

2 Research methodology

2.1 Water biological stability

Biofilm is a group of microorganisms that are linked by trophic dependence from autotrophic and heterotrophic bacteria and fungal toxins and their parallel developing consumers, such as flagellates, ciliates, and creepers [7-12].

Bio-stable water is considered to be non-microbial and does not sustain the growth of microorganisms in the water supply network.

The epidemiological risk caused by the presence of biological membrane in the water supply network is related both to the quality and type of water taken by the water treatment plants and to the efficiency of treatment and disinfection processes [13-17].

Based on the conducted studies, it was found that the intensity of microorganisms growth in the water supply network is mainly determined by the presence of organic compounds, in particular biodegradable dissolved organic carbon (BDOC) and assimilable organic carbon (AOC).

The values of these parameters indicate the content of nutrients that can be easily assimilated by water-borne microorganisms. The AOC concentration in the water entering the network is usually 0.1-9% of dissolved organic carbon (DOC) and about 15-25% of biodegradable dissolved organic carbon (BDOC) [7].

The conventional water purification system based on coagulation, sedimentation, filtration and disinfection eliminates the AOC fraction in less than 30% and total organic carbon (TOC) in 22%, while the system extended with an additional biofiltration by activated carbon deposit removes these fractions up to 60 and even 74% [18-25].

For the risk analysis and assessment, the following threshold values for the biological stability parameters of water were adopted: BDOC < 0.25 gC/m^3, $\Sigma N_{inorganic} < 0.2$ gN/m^3, $PO_4^{3-} < 0.03$ gPO_4^{3-}/m^3, on the basis of [7].

* Corresponding author: kpiet@prz.edu.pl

2.2 Method of risk analysis and assessment of lack of water biological stability

The value of the risk of biological instability of water in the water supply network describes the so-called risk function f(r), defined as the expected value of losses under certain system operating conditions that influence the system's vulnerability (resistance) to threat [3, 26-30].

In the risk analysis, the priority is to identify potential threats. Risk assessment is a comparison of the obtained risk value with the assumed criteria values.

For the analysis of the risk of loss of water biological stability, the following risk definition was adopted:

$$r = E[C \geq C_{gr} = f(S_1, S_2, S_3)] \quad (1)$$

where $E(C \geq C_{gr})$ is the expected value of losses C greater than the accepted limit losses C_{gr}. where S_1 is the criterion corresponding to the content of BDOC, gC/m^3, S_2 is the criterion corresponding to inorganic content $\sum N_{inorganic}$, gN/m^3, S_3 is the criterion corresponding to PO_4^{3-}, gPO_4^{3-}/m^3.

The limit values were proposed for the probability of exceeding the parameters BDOC, $\sum N_{inorganic}$ and PO_4^{3-} based on the criteria contained in the work [7].

Assuming that P_S is the probability of not exceeding the water quality parameters (BDOC, $\sum N_{inorganic}$ and PO_4^{3-}), P_i is the probability of not exceeding the water quality parameters (BDOC, $\sum N_{inorganic}$ and PO_4^{3-}) in a given water sample, P_p is the probability of exceeding the water quality parameters (BDOC, $\sum N_{inorganic}$ and PO_4^{3-}) and P_{pi} is the probability of exceeding the water quality parameters (BDOC, $\sum N_{inorganic}$ and PO_4^{3-}) in a given water sample.

$$P_S = \sum_{i \in E+} P_i \quad (2)$$

$$P_i = \prod_{j=1}^{M} P_{i0} \cdot (1 - P_{i0}) \quad (3)$$

$$P_S = \sum_{i \in E+} P_i \quad (4)$$

The following criteria for the probability of exceeding BDOC, $\sum N_{inorganic}$ and PO_4^{3-} parameters, based on the formulas 2-4 and the work [7]:

- Tolerable risk (R_T) → $C_{gr} = f(S_1, S_2, S_3) = (S_1 = BDOC < 0.25\ gC/m^3) \wedge (S_2 = \sum N_{inorganic} < 0.2\ gN/m^3) \wedge (S_3 = PO_4^{3-} < 0.03\ gPO_4^{3-}/m^3)$
- Controlled risk (R_{CI}) → $C_{gr} = f(S_1, S_2, S_3) = (S_1 = BDOC < 0.25\ gC/m^3) \wedge (S_2 = \sum N_{inorganic} < 0.2\ gN/m^3) \wedge (S_3 = PO_4^{3-} \geq 0.03\ gPO_4^{3-}/m^3)$
- Controlled risk (R_{CII}) → $C_{gr} = f(S_1, S_2, S_3) = (S_1 = BDOC < 0.25\ gC/m^3) \wedge (S_2 = \sum N_{inorganic} \geq 0{,}2\ gN/m^3) \wedge (S_3 = PO_4^{3-} < 0.03\ gPO_4^{3-}/m^3)$
- Controlled risk (R_{CIII}) → $C_{gr} = f(S_1, S_2, S_3) = (S_1 = BDOC \geq 0.25\ gC/m^3) \wedge S_2 = \sum N_{inorganic} < 0.2\ gN/m^3) \wedge (S_3 = PO_4^{3-} < 0.03\ gPO_4^{3-}/m^3$
- Controlled risk (R_{CIV}) → $C_{gr} = f(S_1, S_2, S_3) = (S_1 = BDOC \geq 0.25\ gC/m^3 \wedge S_2 = \sum N_{inorganic} \geq 0.2\ gN/m^3 \wedge S_3 = PO_4^{3-} < 0.03\ gPO_4^{3-}/m^3$
- Controlled risk (R_{CV}) → $C_{gr} = f(S_1, S_2, S_3) = (S_1 = BDOC \geq 0.25\ gC/m^3) \wedge (S_2 = \sum N_{inorganic} < 0.2\ gN/m^3) \wedge (S_3 = PO_4^{3-} \geq 0.03\ gPO_4^{3-}/m^3)$
- Controlled risk (R_{CVI}) → $C_{gr} = f(S_1, S_2, S_3) = (S_1 = BDOC < 0.25\ gC/m^3) \wedge (S_2 = \sum N_{inorganic} \geq 0.2\ gN/m^3) \wedge (S_3 = PO_4^{3-} \geq 0.03\ gPO_4^{3-}/m^3)$
- Unacceptable risk (R_{una}) → $C_{gr} = f(S_1, S_2, S_3) = (S_1 = BDOC \geq 0.25\ gC/m^3) \wedge (S_2 = \sum N_{inorganic} \geq 0.2\ gN/m^3) \wedge (S_3 = PO_4^{3-} \geq 0.03\ gPO_4^{3-}/m^3)$

In Table 1 the values for the probability of exceeding BDOC, $\sum N_{inorganic}$ and PO_4^{3-} parameters were presented.

Table 1. The values for the probability of exceeding BDOC, $\sum N_{inorganic}$ and PO_4^{3-} parameters.

Criteria	PBDOC	$PN_{inorganic}$	PPO_4^{3-}	E(C)
R_T	0.417	0.841	0.99	0.34719
R_{CI}	0.417	0.841	0.01	0.00350
R_{CII}	0.417	0.159	0.99	0.06564
R_{CIII}	0.583	0.841	0.99	0.48540
			$\sum R_{CI-III}$	0.55454
R_{CIV}	0.583	0.159	0.99	0.09177
R_{CV}	0.583	0.841	0.01	0.00490
R_{CVI}	0.417	0.159	0.01	0.00066
			$\sum R_{CIV-VI}$	0.09734
R_{una}	0.583	0.159	0.01	0.00093

In addition, the controlled risk should be adopted if one of the six criteria is met. For risk analysis, the following values of limit losses $E(C<C_{gr})$, related to maintain the biological stability of tap water, were assumed:

- tolerable risk (R_T) - water is biologically stable with the criteria presented in Table 1. The expected value $E(C<C_{gr})$ is expressed by the probability P_{RT} = 0.34719,
- controlled risk (R_C) - there are premises for maintaining the biological stability of water if the parameters shown in Table 1 are maintained. The expected value $E(C<C_{gr})$ with the probability for conditions I-III is $P_{RCI-III}$ = 0.55454, for the rest conditions $P_{RCIV-VI}$ = 0.09734,
- unacceptable risk (R_{una}) - water is biologically unstable at the parameters shown in Table 1. The expected value E (C <Cgr) with the probability P_{una} = 0.00093.

The parameters defining the tolerable risk area indicate that the parameters of water provide the required biological stability in the water supply network (acceptable safety level (SL_T)), whereas the controlled risk means that the water quality parameters indicate the possibility of changes in the chemical stability of water in the water supply network (safety level which requires control and reduction of (SL_C)). The unacceptable risk means the occurrence of such water quality parameters that cause biological instability of water in the water supply network, which may result in the secondary contamination (unacceptable safety level (SL_{UNA})).

3 Analysis of the selected water distribution system

3.1 Characteristics of the research object

The analysis of the underground water treatment system is performed for the city of 80 thousand inhabitants. Water devices for the water intake are SI pumps consisting of 5 drilling wells (Q_{emax} 183 m^3/h) and SII consisting of 22 drilling wells (Q_{emax} = 715 m^3/h). The designed maximum capacity of the water treatment plant (WTP) is 715 m^3/h. The annual average, for example in 2012/2013, was about 265 m^3/h.

The treated water is discharged from the clean water tank to the two water mains through 3 pumps. Two of them have a capacity of 440 m^3/h and the third one the capacity of 280 m^3/h. Their lifting capacity is 58 m. The city's water supply network consists of the 16.8 km main network, 176.6 km of the distributional network. The household connections network consists of 3958 connections and has length of 98.40 km. Approximately 42.70% of the entire water supply network is made of pipes less than 20 years old but more than 11 years old. 90.2 km of the network is made of materials less than 10 years old.

The intake water does not meet sanitary requirements in terms of turbidity, colour, permanganate index, ammonium, iron and manganese. The physicochemical composition of the intake water indicates that there may be difficulties in its treatment. The physical and chemical parameters of the intake water are as follows: turbidity 8-14 NTU, colour Hazen 40-100, permanganate index 11.0-18.1 g O_2/m^3, ammonium ion 1.2-1.98 g NH_4^+/m^3, iron 14.0-44.0 g Fe/m^3, manganese 0.74-2.58 g Mn/m^3, TOC 14-20 g C/m^3, total hardness 4.0-9.4 val/m^3, sulfates 60- 240 g SO_4^{2-}/m^3. The intake water is directed to the collection well (stop time 2-4 hours, depending on the current water production). Then the water is directed to the oxygenation cascade. Just beneath the cascade lies the dosage of chemical oxidant - potassium permanganate and coagulant PAX-18. The next step in water treatment is vertical coagulatory and sedimentation chambers with a contact time of 6-8 hours. To the water after the chambers, which is directed to the horizontal settlers, calcium milk is dosed to raise the pH. The settling time in the settlers is several hours. The next step is the filtration of water in the filters at a rate of 1.5–3 m/h and the final step - disinfection with sodium hypochlorite.

3.2 Physicochemical parameters determination of the treated water

Samples of water were taken twice a month from November 2015 to November 2016. The treated water was taken from the clean water tank prior to the final disinfection process. The modified Kaplan and Newbold method was used to determining the BDOC content by using for this purpose colonized by autochthonous bacteria bioreactor with granular activated carbon.

The physicochemical parameters were determined according to the applicable research procedures: ammonium nitrogen - direct nesserization method according to PN-C-04576-4:1994P, nitric nitrogen (III) - colorimetric method with 4-diazobenzenesulfonic acid according to PN-EN26777:1999), nitrate nitrogen (V) - colorimetric method with sodium salicylate according to PN/C-04576.08, phosphates - spectrometric method with ammonium molybdate according to PN-EN ISO 6878, TOC - TOC analyzer Sievers 5310 C/PN - EN 1484:1999, BDOC – the modified Kaplan and Newbold method [25].

3.3 Risk analysis and assessment

The existing water treatment station very effectively removes iron and manganese. In the treated water there are traces of these impurities. On the other hand, the colour parameter is maintained at 10-15 Hazen, with an acceptable Hazen value of 15. In the discussed example the colour was affected by organic substances - the content of TOC in the treated water was 7-12 g C/m^3, exceeding the value recommended by the WHO - 5 g C/m^3. Other parameters of the treated water are as follows: BDOC 2.27-11.20 g/m^3, $\Sigma N_{inorganic}$ 0.16-1.20 g/m^3 and PO_4^{3-} 0-0.007 g/m^3. The detailed analysis of BDOC, $N_{inorganic}$ and PO_4^{3-} parameters was performed to analyse the risk of lack of biological stability.

Fig. 1 shows the parameters of the treated water and the safety criteria established in Table 1.

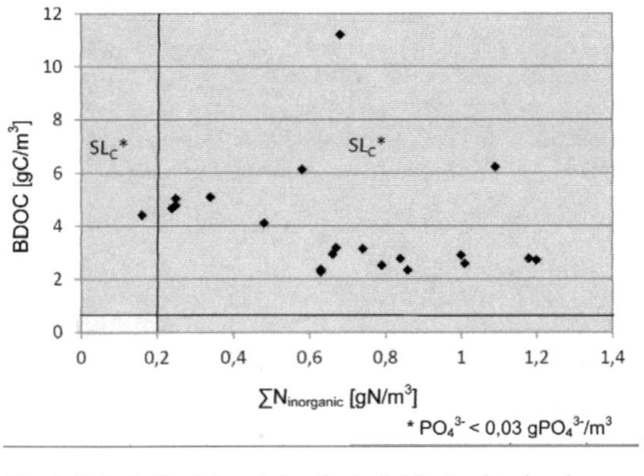

Fig. 1. Criteria for determining the individual safety levels taking into account the parameters of the treated water.

All tested samples are in the range of controlled risk (R_C). Of the 22 tested water samples 21 corresponds to variant IV of controlled risk (R_C):

IV. $C_{gr} = f(S_1, S_2, S_3) = (S_1 = \text{BDOC} \geq 0.25\ gC/m^3) \wedge (S_2 = \Sigma N_{inorganic} \geq 0.2\ gN/m^3) \wedge (S_3 = PO_4^{3-} < 0.03\ gPO_4^{3-}/m^3)$

One sample of water corresponds to variant III of

controlled risk (R_C):III. $C_{gr} = f(S_1, S_2, S_3) = (S_1$ = BDOC ≥ 0.25 gC/m^3) $\wedge$ ($S_2 = \sum N_{inorganic} < 0.2$ gN/m^3) $\wedge$ (S_3 = $PO_4^{3-} < 0.03$ gPO_4^{3-}/m^3)

The controlled risk level (R_C) means that the nature of the water samples determined by BDOC, $\Sigma N_{inorganic}$ and PO_4^{3-} corresponds to a safety level that requires control and reduction (SL_C).

4. Conclusions

Conventional water purification processes i.e. coagulation, sedimentation, filtration and disinfection do not provide effective elimination of biogenic substances and, in particular, BDOC and inorganic nitrogen. Because of the insufficient degree of nutrient removal it is necessary to include to conventional water treatment systems the additional unit processes or to optimize the existing ones.

For the considered technological variant the treated water corresponds to a safety level which requires control and reduction (SL_C) to a level where there are the premises of maintaining the biological water stability. In case of the tolerable risk, the further monitoring and the maintenance of the current level of drinking water quality are recommended.

The controlled risk indicates the need for correction in the water treatment process. In case of receiving the unacceptable level of risk, it is recommended to supplement the technological process, e.g. by using the biofiltration process.

References

1. I. Zimoch, Ochr. Srod. **31** (2009)
2. I. Zimoch, E. Lobos, T. Zaba, *Conf. Proc. Safety and Reliability: Methodology and Applications*. (Boca Raton: CRC Press-Taylor & Francis Group, 2015)
3. M-C. Bessner, V. Gauthier, B. Barbeau, R. Millette, R. Chapleau, M. Prevost, J. Am. Water Works Assoc. **93** (2001)
4. J. Rak, Environ. Prot. Eng. **29** (2003)
5. B. Tchorzewska-Cieslak, J. Rak, *Congress Proc. of Environmental Engineering*, (Lublin, Poland, 2010)
6. J. Królikowska, Rocz. Ochr. Sr. **13** (2011)
7. M. Wolska, *Usuwanie substancji biogennych w technologii oczyszczania wody przeznaczonej do spożycia przez ludzi* (Oficyna Wydawnicza Politechniki Wrocławskiej, Wrocław, 2015)
8. M.J. Lethola, I.T. Miettinen, T. Lampola, A. Hirvonen, T. Vartiainen, P. J. Martikainen, Water Res. **39** (2005)
9. B. Toczyłowska, Tech. Wody **45** (2016)
10. V. Ondrejka Harbulakova, A. Estokova, M. Kovalcikova, Environ. **4** (2017).
11. M. Łebkowska, E. Pajor, A. Rutkowska-Narożniak, M. Kwietniewski, J. Wąsowski, D. Kowalski, Ochr. Srod. **33** (2011)
12. B. Kołwzan, Ochr. Srod. **33** (2011)
13. I. Biedroń, A. Trusz-Zdybek, T. Traczewska, Instal **5** (2013)
14. J. Wingender, H.C. Flemming, Int. J. Hyg. Environ. Health **214** (2011)
15. T.M. Traczewska, M. Sitarska, Ochr. przed korozją **55** (2012)
16. V. Ondrejka Harbuľáková, A. Eštoková, N. Števulová, Nadežda, A. Luptaková, Int. J. Renew. Energy Environ. Eng. **5** (2014)
17. T.M. Traczewska, M. Sitarska, I. Biedroń, *Ekologiczne i techniczne aspekty powstawania biofilmu w wodzie* (Oficyna Wydawnicza Politechniki Wrocławskiej, Wroclaw, 2014)
18. M. Świderska-Bróż. Ochr. Srod. **34** (2012)
19. M. Kabsch-Korbutowicz, *Zaawansowane metody usuwania naturalnych substancji organicznych z wody*. (Komitet Inżynierii Środowiska PAN, Lublin, 2012)
20. A. Pietrzyk, D. Papciak, JCEEA. **63** (2016)
21. D. Papciak, J. Kaleta, A. Puszkarewicz, B. Tchórzewska-Cieślak, J. Ecol. Eng. **17** (2016)
22. M.K. Ramseier, A. Peter, J. Traber, U. Von Gunten, Water Res. **45** (2011)
23. P. Thayanukul, F. Kurisu, I. Kasuga, H. Furumai, Water Res. **47** (2013)
24. Ch. J. Volk, M. W. LeChevallier, J. Am. Water Works Assoc. **92** (2000)
25. L.A. Kaplan, J. D. Newbold, Water Res. **12** (1995)
26. J. Rak, B. Tchórzewska-Cieślak, J. Konbin. **3** (2010)
27. K. Boryczko, B. Tchorzewska-Cieslak, *Conf. Proc. on Environmental Engineering*, (Lublin, Poland, 2012)
28. J. R. Rak, B. Kucharski, Environ. Prot. Eng. **35**, (2009)
29. I. Zimoch, Ochr. Sr. **34** (2012)
30. I. Zimoch, E. Łobos, Environ. Prot. Eng. **38** (2012)

Diversification of water supply

Janusz Rak[1], and *Krzysztof* Boryczko[2,*]

[1] Faculty of Civil and Environmental Engineering and Architecture, 35-959 Rzeszow, Poland

Abstract. The subject of the publication is the presentation of a methodology for determining the degree of diversification of water resources in collective water supply systems (= CWSS). Knowing the number of subsystems for water supply and their share of total water production, it is possible to calculate the dimensionless Pielou index. Similarly, the diversification indicators for networked water tanks (number and volume) and pressure pipelines of the second degree pumping station (number and flowability) can be determined. The work presents the calculation of diversification indices for selected CWSS in Poland. The presented methodology gives the possibility of three-parameter evaluation of settlement units with different water demand and different technical structure.

1 Introduction

Known in economics portfolio diversification one of the most effective methods of reducing investment risk. This concept means the division of the portfolio into different types of investments, among others, in terms of the type of market (e.g. raw materials, currency, shares, bonds), trade (in the case of shares) or geographical coverage of given entities (e.g. shares, shares funds of enterprises from a specific region). In the natural sciences diversification is identified with the biodiversity of fauna and flora in different biosystems [1-4]. Diversification is an important issue in the electricity supply of urban agglomerations [5,6]. Collective water supply system (CWSS) is critical infrastructure which is one of the basic systems for the functioning of urban agglomerations to ensure the security of drinking water supply by public water supply. In the first place CWSS should cover the current and prospective water demand in a technically and economically justified way with respect to natural water resources conservation. Security concise is defined as the ability of a system to protect the interior from external hazards. Analysis and evaluation of both CWSS reliability and security operation indicate the significant role of water resources diversification. Each agglomeration for the proper functioning requires good quality water supply in necessary amount [7]. It is particularly important in crisis or emergency situations. The possibility of water supply from several sources clearly increases the diversification of water supply [8, 9]. If you shut off the water intake or the water intakes the volume of water stored in water supply network tanks plays a key role. An important role is also played by the collective pressure pipelines of the second degree pumping station, whose number and diameter directly affect the reliability of water supply [10-15]. Reducing the water supply or a total lack of water supply can cause threat to water consumers lives or health, as well as financial losses of the recipients and water companies [15-24]. The operator of the CWSS having a high degree of diversification has greater room for manoeuvre in case of crisis, and his decisions do not have to be taken under pressure of time.

The theme of the publication is to present the methodology for determining water supply diversification degree of CWSS. Parameters that have a significant impact on the water supply diversification were proposed:

- number of sub-systems of water supply and their share in the total supply in CWSS,
- number of water tanks and their volume fraction in the whole CWSS,
- number of water mains coming out of the second stage pump with appropriate data rates.

On the basis of these data it is possible to calculate the dimensionless Pielou ratio and assess the degree of diversification of the given collective water supply system, as well as compare even very different water supply systems. The paper presents the calculation of diversification indexes for selected CWSS in Poland.

2 Methods

Evelyn Crystal Pielou was an ecologist by education. She significantly contributed to the development of mathematical ecology, mathematical modelling of natural systems. She is the author of six scientific books on this subject [4, 25-30].

From her work on the assessment of the biodiversity degree of biocoenosis, a measure to assess the degree of diversification of water supply to the city was adopted, determined by the formula:

$$d_P = \frac{d_{SW}}{d_{SW\,max}} \qquad (1)$$

* Corresponding author: kb@prz.edu.pl

where: d_{SW} is the Shannon-Weaver index, calculated according to:

$$d_{SW} = -\sum_{i=1}^{n}(u_i \cdot \ln u_i) \quad (2)$$

and value d_{SWmax} is calculated as:

$$d_{SWmax} = \ln(n) \quad (3)$$

where:
u_i – share of the i-th element in the entirety,
n – number of elements.

Analysing the formulas (1) and (3) it can be concluded that: for n = 1 the index d_P is undefined, for n = 2 the index $d_P = 1.0$.

The analysis of the indexes d_P shows that:
• with the lack or little unevenness of shares of the index d_P obtain values near or equal to 1.0,
• with the lack of significant balance of shares, the rule that the bigger number of n the higher index d_P is not applicable.

It should be noted that for any number n, with the same shares the Pielou index obtains the maximum value of an even diversification of 1.0.

For the assessment of diversification the following parameters were proposed:
Q – parameter associated with the resource of water in the water supply subsystems (WSS) [31],
V– parameter associated with the volume of water in the network water tanks [32, 33],
M – a parameter related to flow ability of pressure pipelines of the second degree pumping station [15, 24].

Water resources in the individual WSS mean the maximum daily production capacity. The parameter V includes the volume of the network tanks, i.e. all the tanks located in the technological line after the second stage pumping stations. The study used the specific flowability (m^3/s) and taking into account the diameters of pressure pipelines of the second degree pumping station, directly correlated them with the flow. The M parameter takes into account the diameters of the collective pressure pipelines of the second degree pumping station and their roughness. The proposed methodology adopted the same roughness of collective pressure pipelines of the second degree pumping station for the tested systems of public water supply. In further studies it will be possible to take into account the material and age of the pipe (the pipe will have a specific value of roughness).

Diversification degree for parameter Q, V, M were determined according to:

$$d_Q = \frac{-\sum_{i=1}^{m}(u_i \cdot \ln u_i)}{\ln m} \quad (4)$$

$$d_V = \frac{-\sum_{j=1}^{n}(u_j \cdot \ln u_j)}{\ln n} \quad (5)$$

$$d_M = \frac{-\sum_{k=1}^{p}(u_k \cdot \ln u_k)}{\ln p} \quad (6)$$

where: d_Q = the diversification index of water resources (intakes) in the CWSS; d_V = the diversification index of the volume of water in the network water tanks; d_M = the diversification index of flowability of collective discharge pipelines of the second stage pumping stations; u_i = share of maximum daily production capacity of the i-th CWSS (m^3/d) in the total maximum daily capacity of water supply to CWSS; u_j = share of the volume of the j-th tank (m^3) in a total volume of water in the network water tanks; u_k = share of the flowability of k-th pressure pipelines of the second degree pumping in a sum of flowability of pressure pipelines of the second degree pumping; m = number of WSS ;n = number of network tanks; p = number of collective discharge pipelines of the second stage pumping stations.

Global diversification degree was calculated according to:

$$d = d_Q + d_V + d_M \quad (7)$$

In the calculations it was assumed that if $u_i = 1.0$, $u_j = 1.0$ or $u_k = 1.0$, the value of the index d_Q, d_V or d_M taken to calculate the index d from the equation (7) is 0.0.

The following standards for the ratio d were adopted:
- lack of diversification $d = 0$
- low diversification $0 < d \le 1.5$
- average diversification $1.5 < d \le 2.0$
- sufficient diversification $2.0 < d \le 2.5$
- very satisfactory diversification $2.5 < d \le 3.0$

3 Results and discussion

Calculations of the assessment of the diversification degree for selected CWSS are presented below.

- Rzeszów

- Q - two intakes:
$u_1=0.43$
$u_2=0.57$
According to (4):

$$d_Q = \frac{-(0.43 \cdot ln0.43 + 0{,}57 \cdot ln0.57)}{ln2} = 0.986$$

- V - twelve tanks:
$u_1=0.018$
$u_2=u_3=0.054$
$u_4=u_5=u_6=u_7=0.088$
$u_8=0.522$
$u_9=u_{10}=u_{11}=u_{12}=0.02$
According to (5):

$$d_V = \frac{-(0.018 \cdot ln0.018 + 2 \cdot 0.054 \cdot ln0.054 + 4 \cdot 0.088 \cdot ln0.088 + 0.552 \cdot ln0.552 + 4 \cdot 0.02 \cdot ln0.02)}{ln12} = 0.743$$

- M - four discharge pipelines of the second stage pumping stations:
$u_1=0.03$
$u_2=u_3=0.05$
$u_4=0.87$
According to (6):

$$d_M = \frac{-(0.03 \cdot ln0.03 + 2 \cdot 0.05 \cdot ln0.05 + 0.87 \cdot ln0.87)}{ln4} = 0.553$$

According to (7):

$d = 0,986 + 0,743 + 0,553 = 2,282$

- sufficient diversification

Fig. 1 shows the values of d_Q, d_V, d_M for analyzed CWSS.

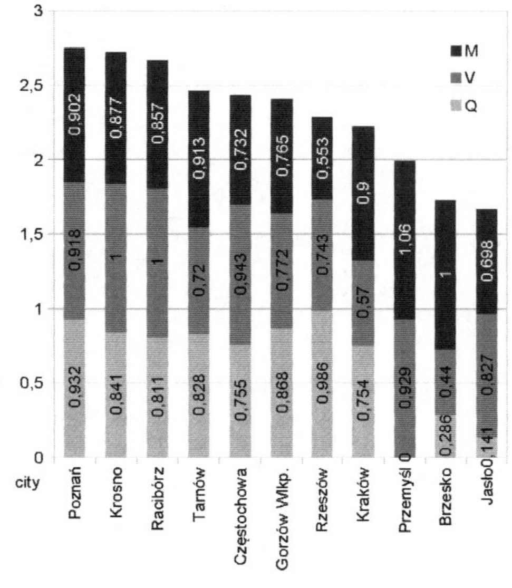

Fig. 1. Values of d_Q, d_V, d_M for analyzed CWSS.

Five out of the analysed CWSS, were classified in the category of sufficient diversification. Highly satisfactory diversification was found for three CWSS, which is due to balanced water resources, volumes and a high value of the index d_M. As for the proposed in the method three parameters, there are always two of them in the given CWSS, i.e. water intakes and discharge pipelines. In addition, the lack of diversification of water resources in the CWSS Przemysl (one intake, $d_Q = 0$) is the cause of the low value of the global index of diversification. With the increase in the number of water intakes, water tanks or collective discharge pipelines of the second stage pumping stations the value of the indexes d_Q, d_V, d_M, grows at balanced shares.

4 Conclusions

The degree of diversification of water supply in the CWSS is greatly affected by:

- number of the WSS, network tanks, collective discharge pipelines of the second stage pumping stations, evenness of distribution of WSS productivity, volume of network tanks, specific flowability of collective discharge pipelines of the second stage pumping stations,
- share of the volume of accumulated water in the size of its collection from the water supply system (the development of criteria and standards is in the process of research).

The CWSS safety is closely related to the diversification of water supply to consumers. It is a desirable situation when during the failure of one of the CWSS elements the recipient of water does not feel any inconvenience related to the operation of the water supply network. The presented method for determining the degree of diversification is a simple tool in the hands of the CWSS operators, showing how the CWSS is prepared to a statutory obligation to supply water.

Dimensionless values of the global diversification index predispose it to analyse the degree of diversification of water supply in different CWSS. They make it possible to compare even very different systems, with one or several intakes, equipped or not equipped with water supply tanks, with different numbers and flowability of collective discharge pipelines of the second stage pumping stations.

The authors are aware that the analyses of the diversification performed in particular cities are result of the technical conditions, the size of water resources in each source, the need for expansion of the CWSS elements. The designers of these systems certainly were not guided by the ability to calculate and assess the degree of diversification because they did not know such concept. We realize that a large number of cities are supplied from a single source (e.g. for Montreal it is a big St. Lawrence river) and the CWSS operate in those cities with satisfactory reliability. The content of the work is a new contribution to the possibility of conscious design of the CWSS expansion, taking into account the diversification as the basis of preliminary reliability.

References

1. R. Van Velzen, N. Wahlberg, M. Sosef, F. Bakke, Biol. J. Linn. Soc. **108**, (2013)
2. M. Hill, Ecology **2** (1973)
3. E. Simpson, Nature **163**, (1949)
4. E. C. Pielou, J. Theor. Biol. **13**, (1966)
5. R. Muller, M. Steinert, S. Teufel, Energ. Policy **36**, (2008)
6. B. Tchórzewska–Cieślak, K. Pietrucha-Urbanik, M. Urbanik, Eksploat. Niezawodn. **18**, (2016)
7. W. Kuczyński, W. Żuchowicki, Rocz. Ochr. Sr. **12**, (2010)
8. S. Biedgunis, M. Smolarkiewicz, P. Podwójci, A. Czapczuk, Rocz. Ochr. Sr. **9**, (2007)
9. M. Dalewska-Kolan, P. Podwójci, Rocz. Ochr. Sr. **11**, (2009)
10. J. Bajer, *Ecomical and reliability criterion for the optymization of the water supply pumping stations designs. in: Environmental Engineering IV* (Taylor & Francis Group, London, 2013)
11. J. Bajer, *Reliability analysis of variant solutions for water pumping stations. in: Environmental Engineering* (Taylor & Francis Group, London, 2007)

12. B. Tchórzewska-Cieślak, Ochr. Sr. **29** (2007)
13. B. Tchórzewska-Cieślak, D. Szpak, Ochr. Sr. **37** (2015)
14. D. Szpak, B. Tchorzewska-Cieslak, *Water producers risk analysis connected with collective water supply system functioning. in: Dependability Engineering and Complex Systems. Advances in Intelligent Systems and Computing,* (Springer, Cham 2015)
15. I. Zimoch, Ochr. Sr. **34** (2012) http://www.os.not.pl/docs/czasopismo/2012/4-2012/Zimoch_4-2012.pdf
16. E. Szymura, I. Zimoch, Przem Chem. **93**, (2014)
17. K. Boryczko, A. Tchorzewska-Cieslak, *Analysis and assessment of the risk of lack of water supply using the EPANET program in: Environmental Engineering IV* (Taylor & Francis Group, London, 2013)
18. B. Tchórzewska-Cieślak, K. Boryczko, M. Eid, *Failure scenarios in water supply system by means of fault tree analysis. in: Advances in Safety, Reliability and Risk Management*, (Taylor & Francis Group, London, 2012)
19. K. Pietrucha-Urbanik, A. Studzinski, Rocz. Ochr. Sr. **18** (2016)
20. B. Kucharski, J. Rak, Environ. Prot. Eng. **35** (2009)
21. B. Tchórzewska-Cieślak, K. Boryczko, I. Piegdoń. *Possibilistic risk analysis of failure in water supply network. The European Safety And Reliability Conference (ESREL)*, (Taylor & Francis Group, London, 2015)
22. K. Boryczko, I. Piegdoń, M. Eid, *Collective water supply systems risk analysis model by means of RENO software. in: Safety, Reliability and Risk Analysis: Beyond the Horizon,* (Taylor & Francis Group, London, 2014)
23. K. Pietrucha-Urbanik, Eng Fail Anal **57**, (2015)
24. I. Zimoch, Ochr. Sr. **30** (2008) http://www.os.not.pl/docs/czasopismo/2008/Zimoch_3-2008.pdf
25. E. C. Pielou, *The interpretation of ecological data: a primer on classification and ordination* (Wiley, New York, 1984)
26. E.C. Pielou, *Biogeography* (Wiley, New York, 1979)
27. E.C. Pielou, *Mathematical ecology* (Wiley, New York, 1977)
28. E.C. Pielou, *Ecological diversity* (Wiley, New York, 1975)
29. E.C. Pielou, *Population and community ecology: principles and methods* (Gordon and Breach, New York, 1974)
30. E.C. Pielou, *Introduction to Mathematical Ecology* (Wiley, New York, 1969)
31. J. Rak, A. Włoch, *Models of level diversification assessment of Water Supply Subsystems. in: Underground Infrastructure of Urban Areas 3,* (Taylor & Francis Group, London, 2015)
32. J. Rak, IJCEE **32** (2015)
33. I. Piegdon, B. Tchorzewska-Cieslak, D. Szpak, *The use of geographical information system in the analysis of risk of failure of water supply network. in: Environmental Engineering V* (Taylor & Francis Group, London, 2017)

Process of hydraulic models calibration

Izabela Zimoch[1,*], *and Ewelina* Bartkiewicz[1]

[1]Silesian University of Technology, Institute of Water and Wastewater Engineering, 44-100 Gliwice, 18a Konarskiego St., Poland

Abstract. Mathematical modelling of the water supply systems (WSS) and water quality changes in the system is a complex and difficult task to solve, it requires an interdisciplinary approach to considering the determinants of WSS work. Prognosis models of the WSS in relation to hydraulic quantities are well known and there are many packages that implement these models. These packages allow you to calculate the flow and pressure in the water distribution system under certain operating conditions. However, to make a hydraulic model a useful tool in the management of water supply systems, a calibration process is required. This process involves estimating model parameters to minimize the difference between model results and actual observations. This is a complex and multi-stage process where the network graph and parameters such as roughness coefficient, pump characteristics, or nodal demands are checked and corrected. The following work contains a complex process of calibration of the actual WSS that supplies water to the customers of the selected part of the Silesian agglomeration.

1 Introduction

A popular trend among water supply companies is the introduction of software to WSS modelling. There are many commercial software that are used to build a model and carry out simulations, such as EPANET, PICCOLO, Mike Net, Water GEMS, and ISYDYW. Increasingly, this software also enables the integration between GIS (Geographic In- formation System), SCADA (Supervisory Control And Data Acquisition) and CIS (Customer Information System) databases, which makes it easier to work on building a WSSs model. Mathematical models reflect operational work of the WSS, thus allowing the analysis of various events occurring in the water distribution subsystem without interfering with real objects. They also support making decisions such as: modernization or extension of water distribution network. Thus, the accuracy of the model must be at a high level. Creating a model is an complex process requiring knowledge of the structure and characteristics of operational work of WSS, and also in the field of modelling. The best solution for such task is its division into individual stages [1]. This approach works well for any modelling project. The most important stages of this process include: data collecting, data verification, testing the model, model calibration and validation, starts analyzes.

During creating hydraulic models, errors may occur, mainly related to the structure and geometry of the network, therefore all stages are closely related to each other creating a loop. If the model reflects the work of WSS, this loop is closed. Fig. 1 shows the action course during the process of creating a hydraulic model. The calibration process plays a main role in these activities, which in fact enables the creation of a hydraulic model and the launching of simulations.

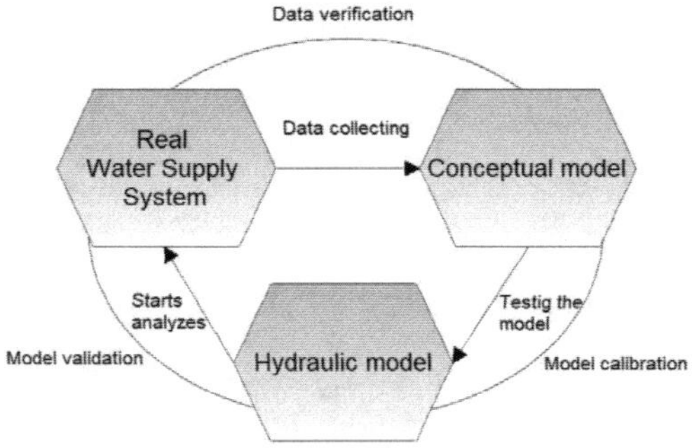

Fig. 1. Diagram of the procedure for creating a hydraulic model.

Calibration is extremely the most difficult and the most important part of modelling, consisting in determining the physical and operational characteristics of the considered WSS. The final stage is to estimate the consistency of the results obtained in the model with the actual field observations. In the case of a large discrepancy between the considered values, the attributes of the objects are again determined until the desired compatibility of the tested parameters is obtained. Calibration, like modelling, can be divided into stages. Seven main steps of the calibration process were determined [1-4]:

a) defining the use of a hydraulic model,
b) initial estimation of the model parameters (conceptual model),

* Corresponding author: izabela.zimoch@polsl.pl

c) collecting calibration data,
d) preliminary assessment of the model's results,
e) macro-calibration,
f) evaluation of simulation parameters after macro-calibration,
g) micro-calibration.

The model can be used to develop various development strategies of the water supply system, simulation of various operational event scenarios, so the first step in the calibration is to specify exactly what the model should be used for. Walski [5] emphasizes that a once calibrated model can't be used for all purposes.

The second step of the calibration is to reconstruct the geometry of the network together with the water supply infrastructure and preliminary estimation of its parameters. The source of network geometry information is numerous databases, which often contain errors or are incomplete. A quick replenishment of these deficiencies is often impossible due to the lack of design documentation or as-built documentation. In these circumstances, the verification of data is time-consuming and requires geodetic field measurements. Currently, water supply companies use in management from GIS systems, which in the case of integration with a hydraulic model, significantly shorten the process of building a network graph. Most properties of objects (diameters of pipes, tanks, hydraulic characteristics of pumps, reducers' settings) included in the model will have a small degree of uncertainty, however one parameter is characterized by high variability, it is the coefficient of pipe roughness [6-10]. Preliminary values of the coefficient of roughness can be taken from the literature. Many researchers and manufacturers [11-13] of pipes have developed tables with the average size of this parameter, taking into account the material, diameter and age of the pipe. However, ultimately this coefficient should be taken directly from field studies.

Another important element of the calibration is a set of calibration data. According to Walski [14], these data can be divided into three categories: "good", "bad" and "useless". "Good" data is defined as a set of information that is collected during the period when the pressure drop in the network is large enough to draw important conclusions from the calibration. "Bad", on the other hand, contain errors due to incorrect measurement of pressure, improper determination of the height of the measuring device or lack of information about which pumps were activated during data collection. And "useless" data is accumulated when the pressure drops and flow velocities are so small that they are within the range of measurement errors. "Bad" data can be rejected immediately, but how can the two other types be distinguished from each other? The author emphasizes that "good" data is collected when the pressure drop in a given period is greater than the measurement error of devices, while the latter fall within these ranges. Unfortunately, during normal operation, especially the oversized WSS, the water flow rates are so small that the pressure drop is small, so most of the data can be incorrectly classified as "useless". However, these data can be used in the model for checking water levels in tanks, adjusting regulators or setting up pumps.

After process of data collection and verification, macro- and micro-calibration can be started. Macro-calibration is the first proper step in the calibration process, during which larger discrepancies (often above 30%) are identified between simulation results and actual data [1]. The reasons for these differences are often related to the export of data from the GIS database, i.e. incorrect pipes attributes, wrong network geometry, incorrect settings of valves and pumps. The only way to resolve these errors is to check the data associated with the network graph. Micro-calibration is the final step of this process, in which pipes roughness and water demands are corrected. This stage can be time-consuming, especially with extensive networks, so it can be divided into two parts: static and quasi dynamic calibration. In the first model parameters are adjusted to pressure and flow rate in connection with steady-state observations, while in dynamic calibration model parameters are adapted to various pressure and flow rates, as well as water level in tanks and flow rates at pump stations.

Referring to the Chaplain [2,3,15], static calibration can be divided into three categories:
1) iterative models (trial and error methods),
2) explicit models (models of hydraulic simulations),
3) implicit models (optimization models).

In the first group, unknown parameters are updated at each step/iteration using pressures to solve non-linear energy continuity equations [6]:

$$G_{L,i} = H_{i,u} - H_{i,d} - \Delta H_i(Q_i) = 0 \tag{1}$$

where:
$H_{i,u}$ - pressure head on the i-th section in the upper node [m];
$H_{i,d}$– pressure head on the i-th section in the lower node [m];
ΔH_i - pressure head change along the section depending on the type of connection [m];
Q_i - flow intensity in the i-th section [m^3/h];
and flows described by linear mass balance equations [6]:

$$G_{J,i} = \sum Q_k + Q_{dem,i} = 0 \tag{2}$$

where:
Q_k - flow rate in the k-th pipes [m^3/h];
$Q_{dem,i}$ –water demand in the node i [m^3/h].
This method requires network graph simplification and is used for small WSS. The second group of models is based on the solution of a wider set of flow continuity equations, using the data obtained from measuring devices. It should be remembered that the number of determined parameters must be equal the number of measurements, and the errors related to the accuracy of measurements are omitted (assuming that all pressure and flows are good). In the last group of implicit models problems are formulated and solved as

optimization problems consisting in minimizing differences between calculated and actual parameters [16].

2 Characteristics of water supply system

The subject of the research is a selected area of the real WSS, supplying water to several urban-industrial agglomerations. This subsystem is supplied by two Water Treatment Plants (WTP "A" and WTP "C"), water well "B" and Pumping Station (PS) "G". The subsystem contains three complexes of Storage Tanks (ST "D", ST "E" and ST "F") with a total capacity of 157 thousand m^3 (Fig. 2).

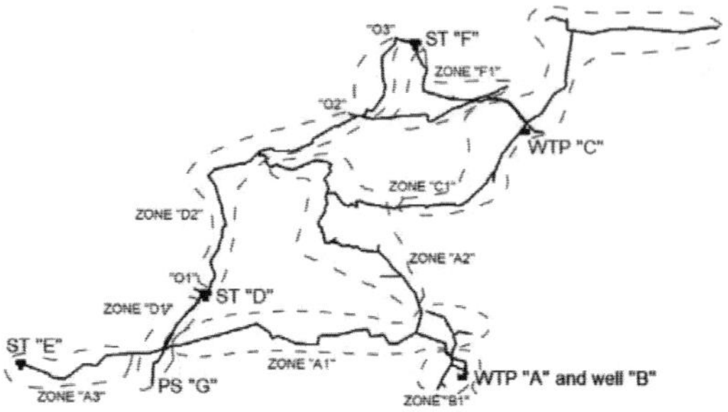

Fig. 2. Scheme of the WSS zone.

In this area there are nodal water demands and three water outflows to another subsystem. On the subsystems map the water outflows are marked with letter "O". Analyzed network is 130 km long with pipes diameters from 79 mm to 1600 mm, and it is made of steel, PE-SDR17 polyethylene and cast iron. The oldest pipelines come from 1929 (steel pipes), and the latest from 2016 (PE-SDR17).

The central point of this subsystem is ST "D" (Fig. 2), to which water flows from two objects (WTP "A" and PS "G") and distribute water in two or three directions (depending on the pressure system). In this case ST "E" are water recipient. WTP "A" (average water production - 39,000 m^3/d) supplies water to the largest area (areas "A1" and "A2"). Well "B" located near SUW "A" captures an average of 725 m^3/d to provide water to residents of nearby area "B1". Reservoirs "F" are supplied from the direction "D2" and distribute water in zone "F1". SUW "C" (average water production - 8000 m^3/d) due to the reduced water production supplies a small area "C1".

3 Calibration of the water supply system model

The model was developed within the framework of NCBiR project POIG.01.03.01-14-034/12. The EPANET 2.0 software was used for the simulation. The calibration process was carried out for data (water production and water demands) from one month (October 2016). Water Distribution System graph was exported from the GIS database, and water demand data was taken from available billing databases. This hydraulic model consists of 871 pipes, 1148 junctions, 302 valves, 11 pumps, 10 tanks and 2 reservoirs. The diurnal curves were determined based on real data collected from the SCADA system for the considered period. In the next step this model was calibrated.

In the first step of model calibration was selection of calibration data. For this purpose were used the filters that reject incorrect measurements of pressure and flow derived from the measurement system on-line. Four data filters were used during export:

• 1 filter skips measurements marked as unreliable by the network explorer,

• 2 filter verifies the data set in terms of the lack of measurements in a given period, if there are fewer than two measurements in a given hour, data is rejected,

• 3 filter verifies data characterized by lack of variability in the analyzed time interval, then the maximum and minimum values are taken into account for the whole period, if these values are the same, then the filter skips them,

• 4 filter refers to measurements characterized by too large variability of measurements, for which a standard deviation is calculated for each hour, if its value exceeds 5% for pressures and 10% for flows, then these values are rejected.

Fig. 3 presents examples of results of using measurement filters. The first digit (85 for flows, 325 for pressures) means the number of measuring points, while the values in brackets are as follows: OK - number of measurements accepted for the calculation, LIST - number of measurements rejected due to the filters described above, NO DATA - number of measurements without a specified value. The last line is the number of all measurements for each point (from several dozen to several hundred measurements per day for one measuring point) and the number of rejected (incorrect) values.

```
;---------------------------------------------------
;WATER FLOWS:85 (50 OK, 31 LIST, 4 NO DATA)

;PRESSURE:325 (254 OK, 41 LIST, 30 NO DATA)

;NO OF MEASUREMENTS:277894 REJECTED:16992 (6.12%)
```

Fig. 3. Example of measurements filters results.

The next stage of the work was the analysis of simulation results. 43 measurement points were tested. Result correlation for the pressure average values was at the level of 99.8%, and flows 98.4%. Despite the results of obtaining a high correlation for average values, the model unfortunately did not reflect the actual work of the network. In these circumstances, three problem areas/objects were identified.

The first object with simulation problem was WT "D", in which the trajectory of tanks filling did not correspond to the measurements (Fig. 4). The reason for this behaviour is the lack of bidirectional flow in the zone "D1" This condition was caused by the failure control of the throttle at reservoirs "G". To improve the calculations, a six-chamber tank was replaced with one with a substitute diameter of 66 m (Fig. 5).

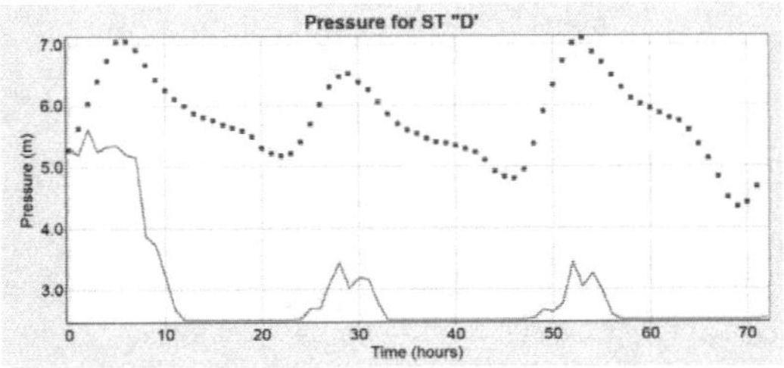

Fig. 4. Trajectory of filling tanks "D" before identifying causes of incompatibility.

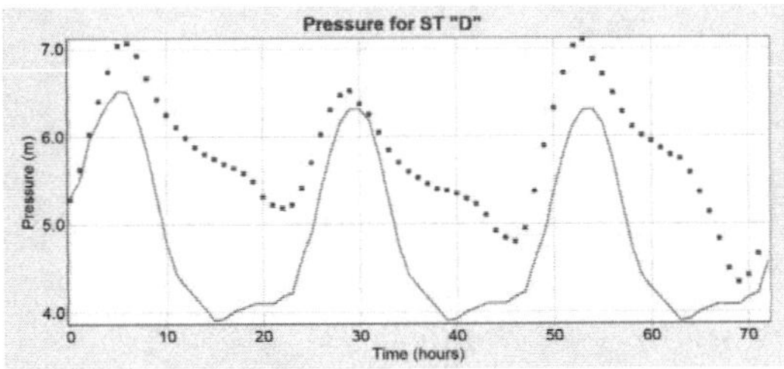

Fig. 5. Trajectory of filling tanks "D" after calibration.

The second case in which the simulation did not correspond to measurements, is zone "A2" supplied by WTP "A" (Fig. 6). The water balance showed that in zone “A2” are discrepancies of water flows of 400 m^3/h between simulation results and field measurements. After the analysis of zones "A2" and "D2" it was found that on the border of these areas there is a closed valve, reducing the water inflow from the area "D2" (Fig. 7).

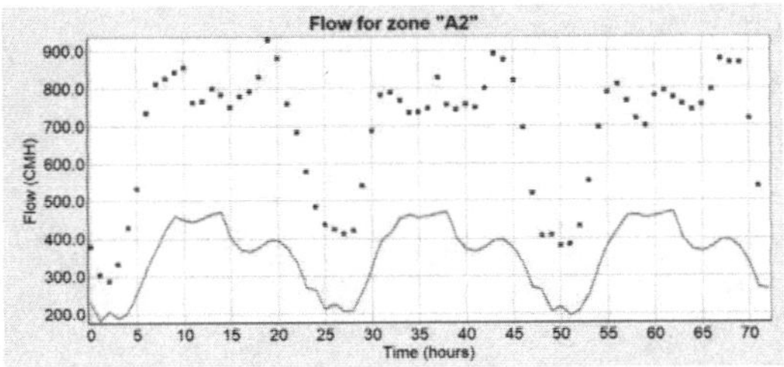

Fig. 6. The water flow in the "A2" zone before identifying causes of incompatibility.

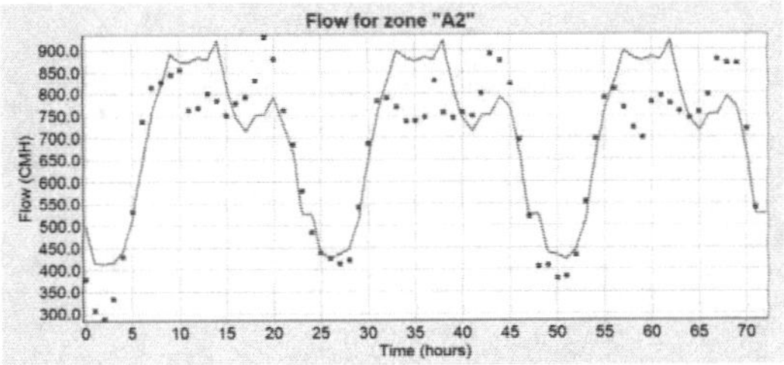

Fig. 7. The water flow in the "A2" zone after calibration.

4 Research results and conclusions

Creating hydraulic models is a complex, multi-stage process during which all elements of the water supply network are verified. Today's IT systems enable the integration of databases and computational programs, thus supporting model building and simulation in real time. However, it should be remembered that an extremely important element of modelling is the calibration process. This process requires not only knowledge in the field of model creation and calibration, but also in the scope of the analyzed network.

Despite the good results of the first simulation, the hydraulic model did not reflect the actual work of the WSS. After analyzing a given zone in problematic areas, it was possible to obtain better results.

The next stage of calibration will be micro-calibration. As part of this work, it will be necessary to determine the pipes roughness coefficients. The data will include data of age and material of pipelines. This will allow to determine the actual/substitute roughness coefficients. This is a very time-consuming task, considering the size of the analyzed system, therefore this step can be solved using the optimization options. In the next stage of micro-calibration verification should be water demands (placements and value) as well as daily water demands patterns.

After carrying out the micro-calibration and validation, the model will become a useful tool to assist in deciding whether to use or modernize the network.

This work was supported withinstatutory fundsBKM-554/RIE-4/2017 and project no BK-286/RIE-4/2017.

References

1. Modelling Special Interest Group, *Water Distribution Modelling Guidelines*, (Water New Zealand, 2009)
2. T.M. Walski, J. Water Resour. Plann. Manage. **109**, 360 (1983)
3. D.A. Savic, Z. Kapelan, P.M.R. Jonkergouw, Urban Water J. **6**, 3 (2009)
4. I. Zimoch, Environ. Prot. Ochr. Sr **30**, 3 (2008) http://www.os.not.pl/docs/czasopismo/2008/Zimoch_3-2008.pdf
5. T. Walski, Proc. AWWA, (1995)
6. L.E. Ormsbee, S. Lingireddy, *Water distribution systems handbook*, (The McGraw-Hill Companies, New York, 2000)
7. M. Kutylowska, Period. Polytech-Civ. **61**, 1 (2017)
8. A. Musz, B. Kowalska, M. Widomski, Ecol. Chem. Eng. S **16**, (2009)
9. M. Kutylowska, M. Orlowska-Szostak, Water Pract. Tech. **11**, 1 (2016)
10. M. Bożym, I. Kłosok-Bazan, M. Wzorek, Pol. J. Environ. Stud. **27**, 3 (2018)
11. Polish Design Standard PN-76/M-34034

12. Z. Siwoń, J. Cieżak, S. Bogaczewicz, Ochr. Sr. **4**, (1998)
13. Fusiotherm company catalog (15.03.2017)
14. T. Walski, J. Am. Water Works Assn. **92**, (2000)
15. Z.S. Kapelan, D.A. Savic, G.A. Walters, J. Water Resour. Plann. Manage. **129**, (2003)
16. I. Zimoch, E. Łobos, Environ. Protect. Eng. **38**, 3 (2012)

Removal of phosphorus compounds from water experience from technological research

Alina Pruss[1,*], and *Paweł* Pruss[2]

[1] Poznan University of Technology, Faculty of Civil and Environmental Engineering, Institute of Environmental Engineering, ul. Berdychowo 4, 60-965 Poznan, Poland
[2] AQUA SA, ul. Kanclerska 28, 60-327 Poznan, Poland

Abstract. The aim of the technological research was to determine the technology which would enable reduction of the concentration of phosphorus compounds in water to the level of 0.02 mg PTot/dm^3. The raw water during the study period was characterized by a pH of 6.66 to 7.62 and a variable concentration of phosphorus compounds ranging from 0.150 to 0.366 mg PTot/dm^3, including phosphates in the range of 0.039 to 0.089 mg PO_4/dm^3. The concentration of chlorides was below 10 mg Cl/dm^3 and sulphates did not exceed 14 mg SO_4/dm^3. Treated water was intended for soil application therefore its quality after treatment could not exceed the limit values set forth in the regulations in force during the study period. The conducted research has shown that the use of the coagulation process based on the ferric coagulant PIX 112 combined with rapid filtration ensures the required quality of the treated water, among other things, through lowering the concentration of PTot to 0.02 mg/dm^3.

1 Introduction

The pollution of surface waters and, more and more often, also groundwaters can be attributed to a high extent to nitrogen and phosphorus compounds. Excessive concentrations of phosphorus in surface water bodies can have a serious negative impact on the natural environment. Phosphorus, just like nitrogen, is one of biogenic elements responsible, among other things, for the deterioration of the quality of water in rivers, streams and lakes, due to eutrophication and blooms. The main sources of phosphorus in surface water include insufficiently treated household and industrial wastewater and agricultural runoff resulting from inappropriate storage or application of natural and mineral fertilisers. In order to make surface water suitable for consumption and industrial and recreational uses and to prevent the occurrence of high phosphorus concentrations, effective water and wastewater treatment technologies must be employed before the effluent is discharged into receivers. The technologies applied in order to overcome the issue should not be too complex but, at the same time, should ensure satisfactory environmental outcomes. The right approach to the subject is to carry out pilot studies in order to select the optimum water treatment technology, define the unit process parameters and determine the size of the necessary investments. The results of technological research help to identify the appropriate process layout and estimated investment and operating costs. Designers receive specific guidelines for drawing up accurate technical documentation and thanks to a well thought-out building design and execution plan the investment can be properly implemented, completed and commissioned, with the guaranteed achievement of the expected technological outcomes [1- 4].

Pursuant to the [5] currently in force in Poland, organophosphorus compounds are classified as substances particularly hazardous to the aquatic environment, causing water pollution which need to be effectively eliminated, while inorganic phosphorus compounds and free phosphorus are classified as substances particularly hazardous to the aquatic environment, causing water pollution whose content needs to be reduced. The target set in the Water Framework Directive [6] was to achieve a good ecological status of surface waters, including rivers, lakes, transitional and coastal waters in the European Union by the end of 2015, among other things, through the reduction of the phosphorus content in surface waters. The Water Framework Directive 2000/60/EC of 23 October, 2000 setting the legislative framework for the community actions in the water policy area is a result of many years of member states' collaborative effort aimed at better protection of the water quality, by way of implementing a common European water policy based on transparent, effective and coherent principles. The directive commits members states to rationally use and protect water resources, in line with the sustainable development guidelines. Consequently, mechanical wastewater treatment plants in Poland were expanded to include biological sections which enable removal of biogenic compounds from wastewater. Another way to prevent phosphorus occurrence in surface waters is to

* Corresponding author: alina.pruss@put.poznan.pl

add reactive substances to the process of their treatment [7-13].

In this article, the authors will present the results of technological research concerning the treatment of water drained from a mine, to be discharged into surrounding lakes. The purpose of the studies was to identify the technology enabling removal of phosphorus from water to reduce its content to 0.02 mg/dm^3. In addition to meeting this parameter, the quality of the treated water had to meet all requirements set forth in the Regulation of the Minister of Environment in force during the study period [14].

2 Methods

Technological research was conducted in March 2014. The pilot station was located at the intake of raw water. The layout of the process equipment used in the technological research is presented in Fig. 1.

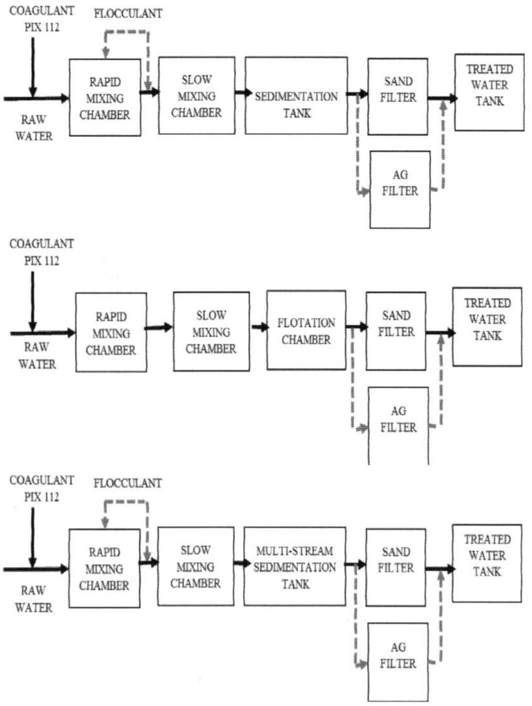

Fig. 1. Process equipment layout during the pilot scale investigation.

Water was sampled from a ditch collecting water from different wells draining the mine grounds. Studies were performed using physical models of coagulation chambers, flotation chambers, a vertical sedimentation tank and a multi-stream sedimentation tank unit. The rate of raw water inflow to the pilot station was 1.5 m^3/h. The coagulant dosage and type were determined in a batch test. The selected coagulant was PIX 112. During the pilot scale process, coagulation in the sedimentation tank was boosted by adding a strongly anionic flocculant.
After separation of the majority of sludge, water was filtered by two connected, parallel rapid filters. The rapid filters were pipes with a diameter of 100 mm and a height of 2,500 mm. In the first filter, a 15 cm gravel support layer was topped with a 100 cm thick layer of quartz sand with a granulation of d_{10}=0.71 mm and d_{60}= 1,25 mm and grain-size uniformity coefficient WR=1.8. In the second filter, a 15 cm gravel support layer was topped with a 100 cm thick layer of AG filter bed with a granulation of d_{10}=0.59 mm, d_{60}=1.65 mm and WR=1.8. Both filters were equipped with an installation enabling control and measurement of the filtration rate and the intensity of backwashing with water. The filtration cycle lasted 48 hours and the filtration rate was 5 m/h. The filters were backwashed with raw water only, with a 50% filter bed expansion.

In order to control the effectiveness of phosphorus compounds removal from water and of the coagulation process, the following water quality parameters and indicators were determined for raw water, water after flotation, water past the vertical sedimentation tank and multi-stream sedimentation tank and the filtered water: temperature, pH, PTot, PO_4, COD (Cr), Nog, N_{NH4}, N_{NO3}, Ca, Mg, Cl, SO_4 and Fe_{Tot}.

3 Results and interpretation

3.1 Selection of the coagulant type and dose

The study evaluated the effectiveness of PIX 112 coagulant used at the rate of 4 and 5 mg Fe/dm^3 and of PIX 112 after alkalization of the water samples with lime water to pH ~ 9. When the water became turbid, PIX 112 was added to pH ~ 7.2. Following coagulation, water samples were collected from selected vessels for analysis and the results obtained are presented in Table 1.

Table 1. Concentrations of total phosphorus and phosphates in the water after coagulation with selected doses of coagulant PIX 112 and calcium hydroxide.

Doses of coagulant mg Fe/dm^3	Concentrations of total phosphorus mg PTot/dm^3	Concentrations of phosphates mg PO_4/dm^3
Raw water	0.180	0.043
PIX 112 – 5 mg Fe/dm^3	0.010	0.020
PIX 112 – 4 mg Fe/dm^3	0.010	0.020
PIX 112 + $Ca(OH)_2$	0.010	0.020

In all water samples after the coagulation process and filtration through a medium-grade filter paper, the total phosphorus concentration was below 0.01 mg PTot/dm^3 and of phosphates below 0.02 mg PO_4/dm^3. The rate of 5 mg Fe/dm^3 was considered the optimum dose of PIX 112 coagulant for use in technological processes. The dose ensured formation of visible sludge flocs across the entire water mass. A similar effect was obtained by using PIX 112 in combination with lime, but the resulting flocs began to sediment during slow mixing and it would not be possible to remove them in the course of the flotation process. Additionally, it was concluded that the use of two reagents and the need to control pH value would make the treatment process more complex. When PIX 112 coagulant was used at the rate of 4 mg Fe/dm^3 sludge flocs formed later, were very small and water was not entirely clear.

3.2 Effectiveness of phosphorus compounds removal in the technological process with a flotation chamber and a rapid filter

Raw water during the study period demonstrated variable pH, ranging from 6.66 to 7.62. Variability was also observed with regard to phosphorus concentrations from 0.150 to 0.366 mg PTot/dm^3, including phosphates in the range from 0.039 to 0.089 mg PO_4/dm^3. Chloride concentrations were below 10 mg Cl/dm^3 and sulphates did not exceed 13.9 mg SO_4/dm^3.

The determined quality parameters and indicators for raw water and water after flotation are presented in Table 2a and 2b, while Fig. 2a and Fig. 2b shows changes in the total phosphorus concentrations in water after flotation and rapid filtration.

Table 2a. Water quality indicators and parameters for raw water.

Water quality indicators and parameters	Unit	Raw water date/sampling time		
		19.03	20.03	28.03
		14.00	8.00	8.00
Temp.	°C	10.40	12.20	10.60
pH	-	6.66	7.06	7.37
Alkalinity	mval/dm^3	-	7.80	-
Turbidity	NTU	12.80	14.50	18.10
Fe $_{Tot}$	mgFe/dm^3	1.14	2.35	2.18
Mn	mgMn/dm^3	0.427	0.416	-
P_{Tot}	mgP/dm^3	0.207	0.366	0.101
PO_4	mgPO_4/dm^3	0.045	0.046	0.066
BOD_5	mgO_2/dm^3	1.00	1.50	1.60
COD	mgO_2/dm^3	9.80	7.70	8.50
N_{Tot}	mgN/dm^3	1.300	1.200	1.230
N_{NH4}	mgN/dm^3	0.686	0.648	0.696
N_{NO3}	mgN/dm^3	0.089	0.067	0.064
Ca	mgCa/dm^3	113.90	114.40	128.70
Mg	mgMg/dm^3	18.10	18.40	20.00
Cl	mgCl/dm^3	9.90	9.90	9.20
SO_4	mgSO_4/dm3	13.90	11.80	11.90

Doses of coagulant PIX 112 = 5.0 – 5.5 mg Fe/dm^3
Doses of floculant = 0.25 – 0.50 mgPAA/ dm^3

Table 2b. Water quality indicators and parameters for water after flotation.

Water quality indicators and parameters	Unit	Water after flotation date/sampling time	
		19.03	20.03
		14.00	8.00
Temp.	°C	10.40	12.50
pH	-	6.72	6.85
Alkalinity	mval/dm^3	-	7.40
Turbidity	NTU	10.50	15.40
Fe $_{Tot}$	mgFe/dm^3	1.91	1.90
Mn	mgMn/dm^3	0.406	0.431
P_{Tot}	mgP/dm^3	0.053	0.051
PO4	mgPO_4/dm^3	0.010	0.020
BOD_5	mgO_2/dm^3	1.10	0.60
COD	mgO_2/dm^3	6.00	7.00
N_{Tot}	mgN/dm^3	1.430	1.380
N_{NH4}	mgN/dm^3	0.688	0.740
N_{NO3}	mgN/dm^3	0.065	0.071
Ca	mgCa/dm^3	116.40	115.90
Mg	mgMg/dm^3	18.60	18.60
Cl	mgCl/dm^3	9.40	10.10
SO_4	mgSO_4/dm3	27.60	28.70

Doses of coagulant PIX 112 = 5.0 – 5.5 mg Fe/dm^3
Doses of flocculant = 0.25 – 0.50 mgPAA/ dm^3

High concentrations of iron compounds and substantial turbidity values were observed in the water past the flotation chamber. The high water turbidity and elevated concentrations of iron compounds indicating an increased suspended solids content accelerated the filtration bed colmatation process and reduced the filtration cycles. Despite that, water after filtration – with minor exceptions – demonstrated acceptable concentrations of phosphorus compounds.

The results presented in Fig. 2 show that after the filtration of water through a quartz sand filter and through an AG filter, the content of phosphorus compounds was equal to 0.02 mg PTot/dm^3 or lower in 77% and 86% of cases, respectively. The average total phosphorus concentration past the quartz sand filter was 0.021 mg PTot/dm^3 and past the AG filter 0.016 mg TP/dm^3.

The studies have proven that it is necessary to use the water filtration process in order to achieve the required concentrations of total phosphorus and phosphates in treated water.

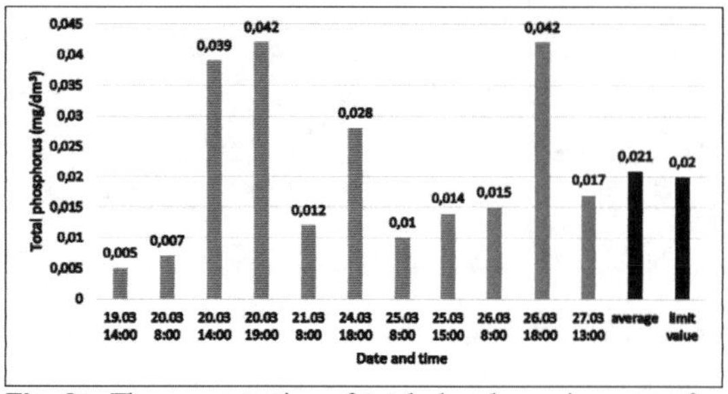

Fig. 2a. The concentration of total phosphorus in water after coagulation with sludge flotation and filtration through a quartz sand filter.

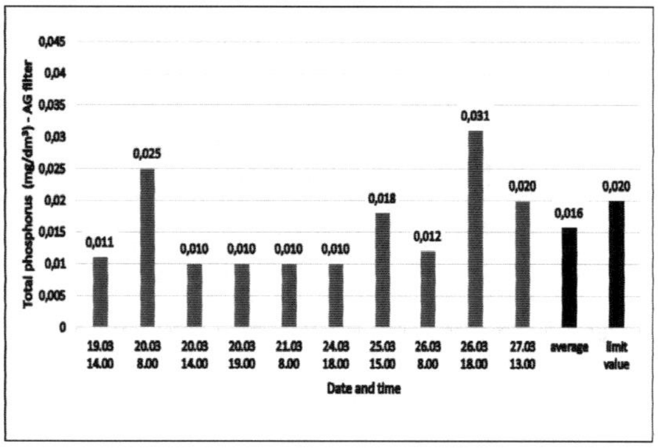

Fig. 2b. The concentration of total phosphorus in water after coagulation of sludge flotation and filtration through AG filter.

3.3 Effectiveness of phosphorus compounds removal from water in the technological process with a vertical sedimentation tank and a rapid filter.

The vertical sedimentation tank removed suspended post-coagulation solids from water more effectively than the flotation chamber, as demonstrated by the lower turbidity and lower iron compounds concentrations.

The turbidity values and concentrations of iron compounds in the water past the flotation chamber and the vertical sedimentation tank are presented in Table 3. It was undeniably an effect of the sedimentation tank's lower capacity versus the flotation chamber by ca. 60%.

Table 3. Turbidity and the iron concentration in water past the flotation chamber and the sedimentation tank.

Water Quality Indicators	Equipment	Average value
Turbidity [NTU]	Flotation chamber	13.4
	Sedimentation tank	4.3
The iron concentration [mgFe/dm^3]	Flotation chamber	4.1
	Sedimentation tank	1.9

The results of coagulation and rapid filtration in the technological process with a vertical sedimentation tank are presented in Fig. 3a and Fig. 3 b and Table 4.

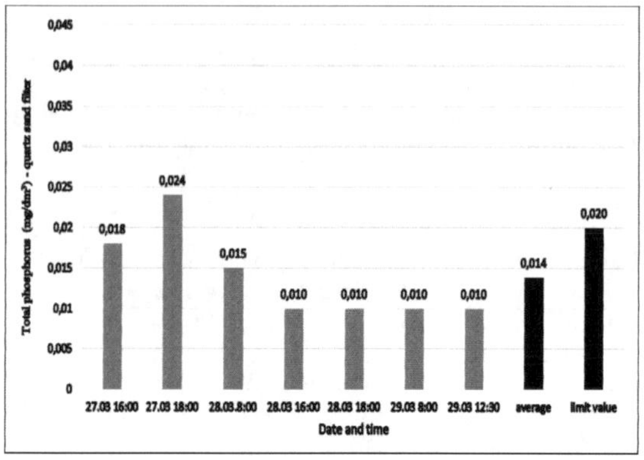

Fig. 3a. Total phosphorus concentration in the water past the sedimentation tank and past the quartz sand filter.

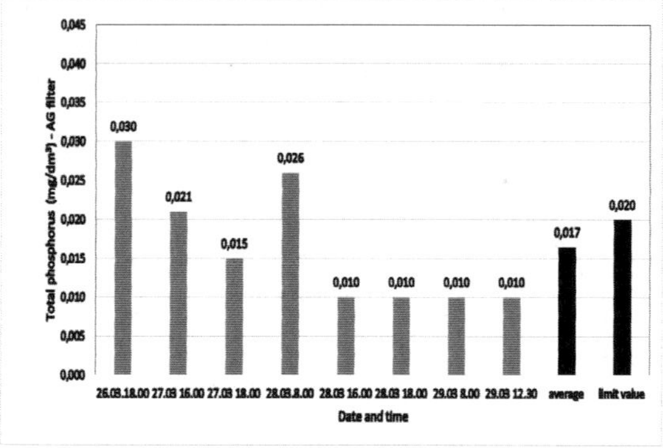

Fig. 3b. Total phosphorus concentration in the water past the sedimentation tank and past the rapid AG filter.

A comparison of Fig. 2 and Fig. 3 shows that as a result of lower concentration of suspended solids past the vertical sedimentation tank (Table 3), the effectiveness of phosphorus compounds removal from water increased. After passing through the quartz sand filter, only one out of seven water samples contained more than 0.02 mg PTot/dm^3, while past the AG filter excessive values were recorded in two cases out of eight.

Worth emphasising is the fact that the average total phosphorus concentrations past both filters were significantly lower than 0.02 mg PTot/dm^3 and amounted to 0.014 mg PTot/dm^3 for the quartz sand filter and to 0.017 mg PTot/dm^3 for the AG filter.

The high effectiveness of phosphorus compounds removal from water in the technological process with the vertical sedimentation tank was related to its limited capacity (0.6 m^3/h), approximately 60% lower than the capacity of the flotation chamber.

The multi-stream sedimentation tank, on the other hand, can be an alternative for the vertical sedimentation tank.

The results of the studies carried out on one multi-stream sedimentation tank are presented in Fig. 4. Given the results presented in Fig. 4, if the effectiveness of suspended solids removal is measured in terms of the maximum turbidity of treated water not exceeding 4 NTU and of the iron concentration close to 1 mg Fe/dm^3

(similar to the average value past the vertical sedimentation tank), the estimated capacity of one unit approximates 13 l/h.

Table 4. Water quality indicators and parameters for the water past the sedimentation tank.

Water quality indicators and parameters	Unit	Water past the sedimentation tank date/sampling time		
		19.03	20.03	28.03
		14.00	8.00	8.00
Temp.	°C	10.4	13.1	-
pH	-	6.6	6.88	-
Alkalinity	mval/dm^3	-	7.40	-
Turbidity	NTU	4.97	9.30	4.34
Fe_{Tot}	mgFe/dm^3	1.95	2.90	1.98
Mn	mgMn/dm^3	0.418	0.420	-
P_{Tot}	mgP/dm^3	0.510	0.092	0.060
PO4	mgPO$_4$/dm^3	0.010	0.011	<0.01
BOD$_5$	mgO$_2$/dm^3	1.20	1.80	1.00
COD	mgO$_2$/dm^3	6.6	8.7	7.9
N_{Tot}	mgN/dm^3	1.290	1.420	1.460
N_{NH4}	mgN/dm^3	0.562	0.730	0.823
N_{NO3}	mgN/dm^3	0.080	0.730	0.072
Ca	mgCa/dm^3	115.4	111.7	122.3
Mg	mgMg/dm^3	18.6	18.5	19.4
Cl	mgCl/dm^3	9.5	9.6	9.6
SO$_4$	mgSO$_4$/dm 3	27.8	30.8	23.4

Doses of coagulant PIX 112 = 5.0 – 5.5 mg Fe/dm^3
Doses of flocculant = 0.25 – 0.50 mgPAA/ dm^3

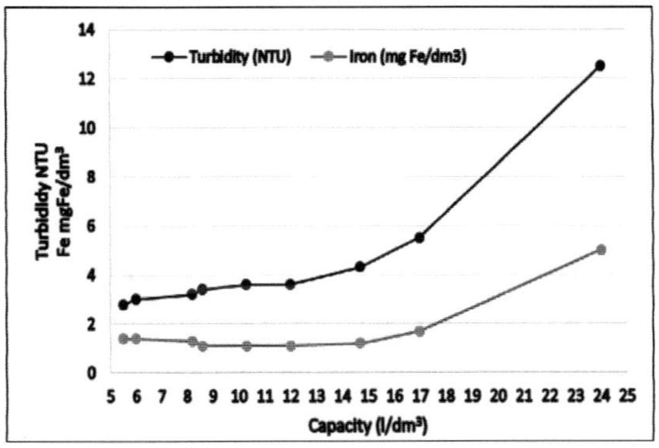

Fig. 4. Turbidity and iron compounds in the water past multi-stream sedimentation tank depending on its capacity.

Based on that value, the capacity of the entire multi-stream sedimentation tank made of pipes with a diameter of 0.05 m can be determined. Due to the inclined position at the angle of 60° and the length of 100 cm, it will consist of 19 x 18 pipes, i.e. the total of 342 pipes. Assuming the unit capacity of 13 l/h, the overall capacity of the whole set will be 4.15 m^3/h.

4 Conclusions

The conducted technological research proved that was possible to treat mine drainage water with the effectiveness ensuring conformance with the requirements of the Regulation of the Minister of Environment of 2006 [14] and, additionally, to achieve reduction of the phosphorus compounds concentration to 0.02 mg PTot/dm^3. The study results remain valid also in the context of the limits set forth in the currently applicable Regulation of the Minister of Environment of 2014 [5].

The technological process in the Water Treatment Plant encompassing coagulation and the PIX 112 iron coagulant, sedimentation and rapid filtration would ensure appropriate quality of treated water.

Due to the low effectiveness of suspended solids removal and relatively high energy consumption of the flotation process, the idea of using flotation chambers was abandoned. Instead, a more effective multi-stream sedimentation tank was proposed.

The filtration of water after the process of coagulation and sedimentation should be carried out using rapid filters with quartz sand or AG filter beds.

The results of the technological research enabled the development of the technological and construction concept of the water treatment plant. Based on the concept, estimated investment and operating costs were estimated. All the documents were used by the Investor in the decision making process regarding the future steps to be taken.

References

1. A. Babatunde, J. Burgess, G. Bertanza, R. Rooney, P. Verlicchi, G. Xu, Water Sci. Technol. **71**, 4 (2015) (DOI: 10.2166/wst.2015.075)
2. A. Pruss A, P. Pruss, Instal **7/8** (2016)
3. A. Pruss, P. Pruss, GWiTS **11** (2010)
4. A. Pruss, Water Sci. Technol. **71**, 4 (2015)
5. Regulation of the Polish Minister of Environment Dz. U. 2014, 1800 (2014)
6. Water Framework Directive 2000/60/EC of 23 October, (2000)
7. A. Kaczmarczyk, A. Bus, *Annals of Warsaw University of Life Sciences – SGGW* **46,** ISSN (Online) 2081-9617. ISSN (Print) 1898 8857.
8. V. Cucarella, G. Renman, J. Environ. Qual. **38** (2009)
9. A. Klimeski, W.J. Chardon, E. Turtola, R. Uusitalo, Agric. Food Sci. **21** (2012)

10. L. Johansson Westholm, Water Res. **40** (2006)
11. C. J. Penn, J.M. Mcgrath. E. Rounds, G. Fox,
12. D. Heeren, J. Environ. Qual. **41** (2012)
13. T. Kirkkala, A.M. Ventela, M. Tarvainen, J. Environ. Qual. **41** (2012)
14. K. Izydorczyk, W. Frątczak, A. Drobniewska, E. Cichowicz, D. Michalska-Hejduk, R. Gross, M. Zalewski, Ecohydrol. Hydrobiol. **13** (2013)
15. Regulation of the Polish Minister of Environment of 24 July 2006 Dz. U. 2006, 137, 984 (2006)

Application of the TEVA-SPOT in designing the monitoring of water networks

Alicja Bałut[1,*], and *Andrzej* Urbaniak[2]

[1]Insitute of Environmental Engineering, Poznań University of Technology, ul. Berdychowo 4, 60-151, Poznań, Poland
[2]Institute of Computing Science, Poznań University of Technology, ul. Piotrowo 2, 60-965, Poznań, Poland

Abstract. TEVA-SPOT software was released in 2008 as a freeware dedicated to water and sewer utilities. It enables modelling of various of measurement points, and built-in tools to conduct more in-depth examinations. This paper presents selected possibilities this program grants in the context of emergencies, terrorist attacks or if it is used as an early warning system. It had gone through a hands-on examination and its results- advantages, difficulties and limitations have as well been described here.

1 Introduction

Consideration of intentional threats that are introduced into a water network on purpose should consider the fact that they may appear at any given spot. Recently it has been raised that water poisoning may be one of terrorists' top 'tactics' due to its relative low cost and, on other end, significant financial, social and psychological losses [1].

Our motivation behind this issue is modern IT solutions dedicated to the management of water and sewage utilities, especially in the area of water networks' hydraulic modelling and early warning systems. Companies even if equipped in expanded platforms and dedicated solutions rarely utilize all their features. An often met problem of water and sewage utilities is lack of integration between geographic information systems, hydraulic modelling, monitoring, SCADA (Supervisory Control And Data Acquisition) system or accidents' recording across an entire company (water intake, filtration, finally distribution). This has been proven to be an issue of high importance, often the one that in case of critical situations decides on the success of the implementation of an integrated management system. It may turn out to be indispensable in critical situations.

Whilst designing an early warning system or a water-sewage monitoring one a broad spectrum of applications of collected data should be considered – not only in order to support their management but primarily to guarantee a fast response and action in case of a threat – a terrorists' attach – even if at present it does not seem likely.

2 Characteristics of the TEVA-SPOT program

TEVA-SPOT (Threat Ensemble Vulnerability Assessment-Sensor Placement Optimization Tool) is the result of research made by National Homeland Security Research Center (NHSRC) in the US that has been realizing scientific research programs since 2002 oriented at i.e. network infrastructure security of United States Environmental Protection Agency (US-EPA). This has been led by researchers from US-EPA, Sandia National Laboratories, Argonne National Laboratory and University of Cincinnati [2]. Over the last several dozens of years US-EPA with support of NHSRC and other research centres has released a range of IT tools dedicated to environment protection, including, i.e. the following programs: CANARY, EPANET –MSX and TEVA-SPOT [3]. TEVA-SPOT was released in 2008 as a freeware; it was dedicated to water and sewage utilities. Its user interface is available only in English. It enables modelling of various scenarios of events that originate from contaminations spreading in water networks. It supports analysis of such events, localization of measurement point and more detailed research addressed in the following chapters of this paper.

2.1 Structure of the program

Functional structure of TEVA-SPOT is based on three key modules:

- First module enables various simulations of incidents in the system that is being designed (described as 'ensemble'). Classification of a single incident may depend on the type of contamination, its amount (dose of the substance), with pointed spot of introduction or contamination, registered time of insertion or detection in a water network.
- Second module is dedicated to forecasting potential effects of detected contamination from the perspective of public health and water network's security.
- Third module runs on multi-criteria optimization methods in order to determine optimal localization of

* Corresponding author: alicja.balut@put.poznan.pl

measurement point whilst minimizing threats for public security and costs and costs [3].
Table 1 presents a set of data that are indispensable as input criteria for calculations performed by TEVA-SPOT. This application can either be used as an add-on, a toolkit for EPANET or as a standalone program with its own user interface that is interfaced with EPANET and utilizes an existing hydraulic model of a water network. The performance of TEVA-SPOT is based on the probability analysis of multiple possible contamination events.

Table 1. Information and data required to design the localization of measurement points using TEVA-SPOT [3].

Data type	Description
Model of a water network	A hydraulic model, a .inp file that performs correctly, where simulations within the 7-10 days can be executed, calibrated.
Characteristics of measuring equipment	Type of equipment, way of probing, ranges of operation, existing system of threats detection, installation cost.
Designing potential threats	Data depicting threat that needs to be detected: a list of potential contaminants, their behaviour, way of propagation in the network, reaction to water.
Measurements	Substance detection time, area and precise localisation, set of people at risk.
Reaction and feedback from the water and sewage utilities	Reaction plan: total time required to limit further exposure of public health at risk.
Potential sites for measurement point	List of measuring equipment that can be installed, based on nodes of the hydraulic model.

The procedure starts with an upload of the file with the hydraulic model. Next, scenarios of possible events and their range, dependent on multiple hydraulic and quality parameters have to be manually defined in the program. Hydraulic calculations are executed in the next step where results derive directly from EPANET's computational engine. These data, saved as files, form the basis for an estimation of an influence on public health and an assessment of an influence of a selected threat on the performance of an analysed water distribution network. The following part of this paper present methodology that has been adopted as well as the procedure how calculations are performed. Data declaration method has been presented at Fig. 1.

3 The potential of TEVA-SPOT from the perspective of analyses it executes

The structure of TEVA-SPOT is sophisticated and based on modules where each is accountable for a set of functions. Every module reads multiple parameters that have to be defined.

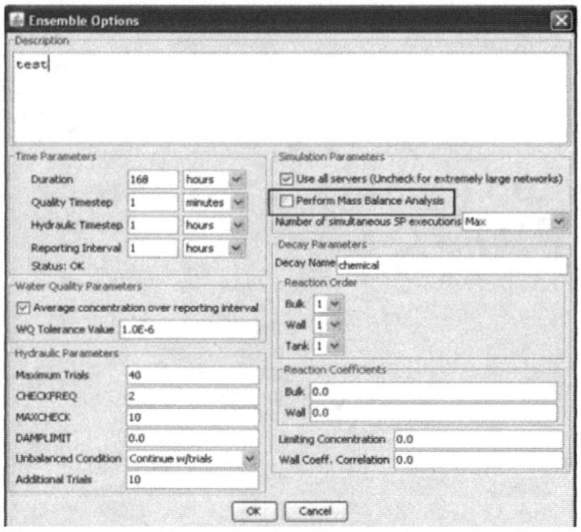

Fig. 1. A screenshot from TEVA-SPOT that presents a view to define basic parameters for threat calculations [4].

Only after all of them that are required to execute calculations have been confirmed, the status bar with available analyzes turns red and they can be executed (Fig. 2).

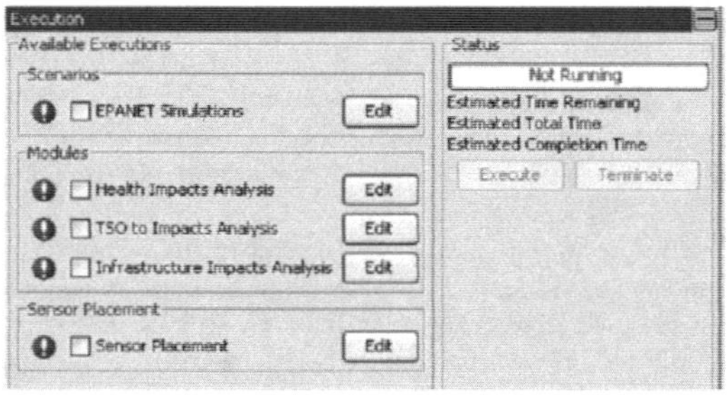

Fig. 2. A screenshot from TEVA-SPOT and the calculation panel for all modules [5].

3.1 Analysis of threat's influence on a water network

Thanks to the IIA module (Infrastructure Impact Analysis) TEVA-SPOT enables analyses of scenarios – influence of threats on network's infrastructure. For every localisation of the detected threat the sum of sections is calculated where concentration of the contamination exceeds the maximum value threshold that had been declared in mg/L. The IIA module supports calculations of sections' length and planning of the sequence of network's rinsing.

3.2 Analysis of threat's influence on a public health

The HIA module (Health Impact Analysis) makes it possible to assess effects of threats on public health, with respect to contamination detected in individual nodes of the model. HIA can also serve as the platform of

 https://doi.org/10.1051/e3sconf/20185900009

detailed information, provided to it beforehand, of analyzed substances and their influence on public health.

At the beginning of the calculation process each threat has to be classified into one of three groups: chemical, toxic, biological. Then, dependent upon the type of the threat a series of factors has to be declared, i.e.: lethal dose (LD) – contaminant's concentration that results in the death of 50% of the population; infectious dose (ID) – sickness or infection of its 50%.

Prior to the initiation of calculations average statistical data of a consumer have to be declared: their weight and daily consumption (default value 1.41 litre). Additionally in case of biological threats a degree of delay has to be stated, being timespan between threat's detection and infection. Fig. 3 presents a sample view with these data.

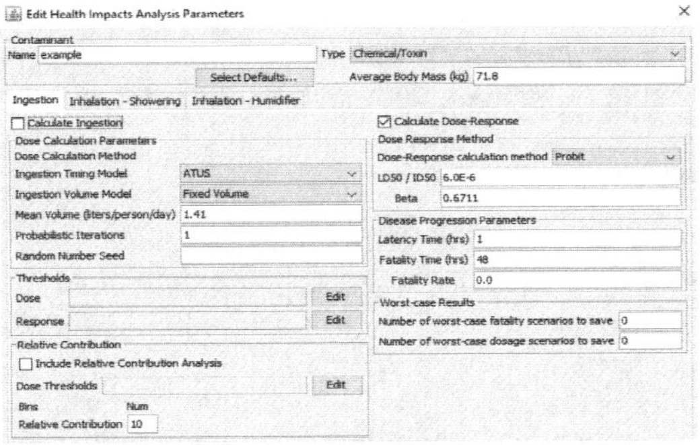

Fig. 3. Data required to run HIA analysis in TEVA-SPOT [5].

To complete the analysis of threat's influence on public health it is required to assign a number of inhabitants consuming water to each node (exposed to consumption of this substance). These actions are undertaken via the 'population analysis' module. Calculation if the population is based on water consumption values, data from billing systems or individual consumption of water, calculated per property [5].

3.3 Analysis of the localisation of measurement points

Analysis of the measurement point localisation is a classical optimisation issue where multiple criteria, limitations and solutions are possible. Choice of these localisation spots has been divided in TEVA-SPOT into a few stages: a) simulation of contamination, b) its influence on public health and network infrastructure, c) multi-criteria optimization that is based on an assignment of weights to criteria. It is possible to simultaneously analyze a few events. Received input data are EPANET results of calculations on hydraulic models as well as files with results of analysis carried within HIA and IIA modules [5].

Localisation of measurement point should take into account users' preferences and ability to install measuring devices [2]. A parameter of significance is assigned to each localization or its preference related to a few analyzed points (perspective of greater probability of pollution). So called expert knowledge, ranking methods (preferences of analyzed spots) and methods of multi-criteria optimization have been applied into this module. TEVA-SPOT gives an option to choose a calculation method [6]:

- Mixed-Integer Programming (MIP) - method that when run for larger networks may seriously consume internal memory.
- Greedy Randomized Adaptive Search Procedure (GRASP) that applies heuristic algorithms. It is efficient when used in case of larger networks yet it demands the problem to be reduced beforehand via shortened list of potential localisations as well as through limitation of simultaneously simulated network threats that is realized during models' skeletisation.
- Lagrange's method of extended multipliers (LAG) – although this method is faster than GRASP, received solutions – precise localisations are not as close to optimal solutions as they are when heuristic methods were applied. This method is used in case of extensive water networks or when available memory is too limited for the GRASP method.

4 TEVA-SPOT in Practice

Application of TEVA-SPOT with a graphical user interface requires earlier installation of two minor programs that support Python as the programming language plus files with JAVA add-ons.

4.1 Ways of presenting results

The TEVA-SPOT program enables creation of various maps, tables and charts that present results of threats' influence calculation on public health as well optimization of the localization of measurement point [7].

Key elements have been listed below: a) maps with estimated fatality at areas impacted by the threat, b) maps of greatest influence on public health, expressed as a 'C' numerical value for the contaminant, c) a table with all water network objects listed by the 'C' value that presents an influence of this threat on all items, d) population maps for all network nodes, e) map containing proposed localisations of measurement point.

Result data can be exported into .mdb/.xls files and when they are needed they can be imported into GIS systems.

4.2 Advantages and drawbacks

For this type of an application, based on calculations realized in the HIA, IIA and RA modules of TEVA-SPOT for a selected theoretical threat, the following list of advantages and limitations have been observed. Testing has been realized at a sample mid-size hydraulic model that had been provided by the software's manufacturer, and a few own similar models. Advantages and drawbacks that have been listed in Table 2 represent the practical aspect of TEVA-SPOT's application and an assessment of user interface with special focus on threat scenarios modelling. First implementation of the program in Poland to localize measurement point is already foreseen. This would give chances for its enhanced assessment from the perspective of a real hydraulic model of a medium-size town in Poland.

Table 2. Advantages and limitations in applying the TEVA-SPOT program.

Advantages	Limitations
Program enables results' assessment of an event, expressed ie. in a number of injured persons and length of networks' sections that require rinsing.	Quality of received data depends on model's quality and its calibration. It is advised to additionally include its representativeness over a year (suggested analysis of two models that include average data for summer and winter seasons.
The authors have underlined that the computational engine is fully scalable what reduces required disc space and operations' runtime to minimum [8].	Lack of any single parameter in the module makes it impossible to proceed with calculations. For models that contain above 10.000 nodes and where a few scenarios have been built calculations can take from over a dozen to a few dozen minutes.
Each module makes it possible to focus on the modelling method and data values.	Use of the „Execution panel" view is not handy and does not always work as expected. Authors have pointed out cases where issues can occur.
Integrity with the EPANET program	Lack of settings' convergence for both applications may lead to errors in calculations performed in TEVA-SPOT. Lack of a direct connection with a GIS platform that provided results to be displayed by crisis management services.
	Threats' definition is difficult due to a lack of research on parameters that characterize substances. This task requires substantial expert knowledge and includes not only details relevant to seasons yet as well local conditions of sites where these substances could be introduced.

5 Conclusions

Solution that has been presented in this paper discusses the oldest IT application that enables simulations of threats and assessment of their impact on water network's integrity. It definitely is a vital supplement to early warning systems and it does contain methodology what actions should be undertaken to design localisation of measurement point in a monitoring system. This article as well addresses potential difficulties, especially for readers who have never before had any experiences with water networks' modelling in EPANET.

According to the authors of TEVA-SPOT, U.S. Environmental Protection Agency has already initiated co-operation with producers of commercial IT solutions aiming at the deployment of analyses discussed in this paper into basic tools of water networks' hydraulic modelling [8].

Limitations that have been presented, that focus on methods of the definition of threats and threshold values, point at the sound requirement to continue research on specific substances' behaviour in various types of water networks, also considering seasoning and individual features of a given one.

Issues that relate to application of IT tools into modelling and design of a monitoring system, that this article presents, are an example of effective use of software interfaced with EPANET. Other cases have been described in our earlier article [9].

References

1. G. Jelonek, Wod. i Kan. **3**, 157 (2017)
2. https://www.epa.gov/aboutepa/about-national-homeland-security-research-center-nhsrc [Updated: 3, 2017]
3. Clark R.M., Hakim S., Ostfeld A., *Handbook of Water and Wastewater Systems*, (Protection Springer, USA, 2011)
4. EPA/600/R-08/147, US-EPA by Janke, R. Threat Ensemble Vulnerability Assessment- Sensor Placement Optimization Tool (TEVA-SPOT) Graphical User Interface User's Manual, NHSRC, Cincinnati, USA, 2012
5. EPA/600/R-09/141. Sensor Network Design for Drinking Water Contamination Warning Systems, A compendium of Research Result and Case Studies Using the TEVA-SPOT Software, NHSRC, Sandia National Laboratories, USA, 2010
6. A. Bałut, A. Urbaniak, Gaz Woda **9** (2013)
7. EPA/600/R-14/433, US-EPA by R. Janke, Application of TEVA-SPOT for Prioritizing Security Enhancements at Utility Facilities and for Protection of Critical Facilities, USA, Nov. 2014
8. R. Janke, TEVA-SPOT: Designing Contamination Warning Systems, Michigan Department of Environmental Quality, NHSRC/US EPA, USA, no. 9
9. A. Bałut, A. Urbaniak, *Proc. of ICCC' 2016, Ivo Petráš – Igor Podlubny – Ján Kačur (eds.)*, (Tatranska Lomnica, Slovakia, 2016)

The quality of groundwater in rural areas and its impact on the functioning of local water intakes

Robert Nowak[1,*], and *Anna* Imperowicz-Pawlaczyk[1]

[1] Department of Environmental Technologies and Bioanalitycs, Faculty of Civil Engineering, Environmental and Geodetic Sciences, Technical University of Koszalin, Sniadeckich 2, 75-453 Koszalin, Poland

Abstract. The main purpose of this paper is to indicate the impact of changes in the quality of groundwater intakes on the functioning of rural water supply systems. In particular, the important factors for the groundwater quality in rural areas have been presented. Results of research on the quality parameters of groundwater, obtained in 2012 ÷ 2016 for selected small water intakes in north-western Poland, were presented. The quality of groundwater in the subsequent analyzed years was deteriorating. There was an increase in the content of iron and manganese compounds, ammonium ions, as well as the presence of bacteriological contaminants in test water samples. For these reasons, problems with providing the receivers with water of appropriate quality parameters have been reported in described rural intakes. The paper is an attempt to show the relationship between the type and magnitude of changes in the quality of groundwater and the efficiency of the treatment plant. In addition, pro-quality activities undertaken by local water companies were emphasized.

1 Introduction

The quality of tap water depends on several factors. The source of water intake, method and the condition of the intake, as well as the efficiency of the water treatment process, should all be taken into consideration. The state of the whole water distribution system is also important. The study of the composition and properties of water intended for drinking is now a routine task for water companies [1].

Over the last decades, more and more villages have been using water supplies. Unfortunately, the construction of collective sewerage and treatment systems has not kept pace with the construction of water supply systems [2]. In rural areas, sewage is generally directed to cesspits, but also to rivers and ponds, often directly to the aerating zone. This results in contamination of the receiving water and land. Sometimes this contamination makes it impossible to use local water resources [3,4]. Land use also affects the qualitative and quantitative status of groundwater resources. Point sources of hazardous chemicals are particularly important [5]. For these reasons, water management in rural areas should be related to the protection of the quality and quantity of water resources. In this area, there is a need for the development of scientific research. This research should aim to explain the impact of farming activities, biological progresses, agro-technical methods, land use, etc. on local groundwater and surface water resources [6]. Intensive and long-term investment in the resources and protection of groundwater quality are also urgently needed [7].

2 Factors that shape groundwater quality in rural areas

In the case of residents of rural areas, the drinking water supply is most often provided through small systems, based on groundwater intakes. In such systems, cases of supplying water with unstable quality parameters are more frequently reported. These problems are often associated with the state of local groundwater resources.

The deteriorating quality of groundwater may have natural causes, but are mainly due to anthropogenic factors. The release and migration of pollutants from surface sources (in particular, cultivation and fertilization of soil and animal breeding) and on-farm activities in rural areas should be mentioned here [8].

For these reasons, aquifers are susceptible to landfill and waste water discharges due to the development of housing and the intensification of agriculture. Important factors for the quality of groundwater are considered: the management of organic and mineral fertilizers and compliance with sanitary regulations concerning the establishment of direct protection zones around water intakes. Problems in this area are present both in Poland and in many other regions of the world [9,10,11]. It should be noted that as a result of administrative decisions taken in Poland in 2010, zones of protection of groundwater intakes have been lost, resulting in the degradation of water resources [12].

The causes of deterioration of groundwater quality may be due to changes in the water circulation system, hydrogeochemical changes and the impact of pollutants. Increases in Fe and Mn concentrations may be due to

* Corresponding author: robert.nowak@tu.koszalin.pl

changes in the water circulation system caused by the mixing of waters from various hydrogeochemical environments (often ammonium nitrate levels, for example, may also increase). The greatest changes in water quality occur as a result of the lowering of the water table during the operation and activation of oxidation processes of sulfides and organic substances. Impacts of anthropogenic impurities are most likely to occur in open and poorly insulated aquifers [13].

An important factor influencing water quality in a given area can be hydrogeochemical changes in the conditions where water is abstracted. Under such conditions, increased levels of nitrates and sulphates can occur in water. Differentiated distribution of nitrate concentrations in the water from rural wells is a result of land development. Sometimes, as a result of hydrogeochemical changes, the chlorine concentration, which is an indicator of anthropogenic pollutants, may also increase [8]. Chloride concentration is highest in waters exposed to pollution from domestic sources. The lowest concentrations are recorded in water in the least populated and non-agricultural areas. Chlorides, by their conservative properties, allow the tracking of the speed and direction of movement of both the water and the substances present therein. However, in the studies on water quality in rural areas, chloride concentrations are often not taken into account. The potential of the data received to identify the sources and fates of substances that negatively affect water quality is also not used [14].

There may be significant degradation in the quality of groundwater due to the impact of pollutants, both large and small, linear and point. Potential foci of pollutants may be located in areas of intensive farming, in areas without collective sewerage and in the vicinity of wastewater discharges. This is also true in areas in which are situated breeding farms, factories and petrol stations. The quality of water can also be affected by landfills [15]. Among the pollutants of groundwater, organic pollutants, including pesticides, should be mentioned. Groundwater susceptibility to these types of impurities depends on natural factors and the properties of the pesticides [16].

Nitrogen compounds, in particular ammonium and nitrates, are basic indicators of the quality of groundwater. The origin of ammonium ions in water can be both natural and anthropogenic. The requirement for low concentrations of these ions in drinking water is primarily due to health reasons. It should be noted that excessive amounts thereof can also lead to secondary pollution of the water (through oxygen consumption and the development of anaerobic bacteria). Detection of ammonium ions in groundwater, where these ions have not previously been identified, most often means contamination of these waters with sewage. This applies particularly to areas of intensive livestock farming, agricultural crops or landfills. Counteracting and reducing the impact of point sources of pollution (mainly those associated with animal production) on rural areas must be associated with monitoring these contaminants [17]. It should be noted that increased concentrations of fertilizer components can endure in contaminated water for a long time, even if commercial farming has been abandoned and farming or breeding is only carried out for personal use [18].

Near point contamination sources, there is a significant variation of the vertical water chemistry, especially in the first (subsurface) aquifer. Identification of processes that shape the subsurface aquifer chemistry may be based on multivariate analysis of the factor. The analysis should take into account a number of data: water chemistry, hydrogeological conditions and forms of land use [19]. In the migration of pollutants, the thickness of the soil layer and the thickness of the aeration zone play a significant role [20,21]. The large vertical variation of the chemical composition of groundwater in the first aquifer has been demonstrated in hydrogeological research [22]. The removal of chemicals from the aeration zone has been observed, particularly with an increase in the depth of sampling the concentration of nitrates, nitrites, sodium and potassium has decreased. The inverse dependence was noted for the concentration of silica, iron and manganese, i.e., ions, penetrating into groundwater due to long-term interaction: the rock-groundwater. In water in rural areas, as opposed to sewerage areas, the concentration of nitrate nitrogen was higher than the concentration of nitrite nitrogen, which occurred in each aquifer. This relationship was observed in each aquifer and indicates that there are conditions that allow a faster transition of nitrogen compounds to oxidized forms. This occurrence may be augmented by other sources of pollution, for example, farming (agricultural fertilization, animal breeding) or poorly developed sewerage infrastructure [21]. Among the factors that are important for the flow of pollutants, the most important should be considered to be the type of production and its intensity and the conditions of land use. Local hydrological and hydrodynamic conditions and physiographic elements that condition infiltration are also important [23]. The spatial distribution of water hazards can be characterized by a high variability for a particular area. Water security at certain times can be influenced by different factors: the distribution of surface water supply, dehydration of groundwater, sanitation, irrigation and the properties of the water [24].

Shallow groundwater, due to its susceptibility to contamination and poor isolation from the penetration of pollutants from the ground surface, may be an indicator of the agriculture impact on the environment [25]. It should be noted that the impact of agricultural activity on the composition of groundwater shows high spatial variability [26]. In shallow groundwater in rural areas, particular attention is paid to the exceedance of drinking water values of nitrate nitrogen concentrations. This happens in wells located near to farm buildings [27]. Nitrate contamination is then the result of incorrect sewage management, including inadequately protected slurry on rural farms [20]. Inadequate water quality in poorly secured wells in non-urbanized areas may also be manifested by excess levels of iron and manganese compounds [28]. Water of the appropriate quality is most often obtained by exploiting deep wells which are appropriately located and isolated from potential pollutants [10]. Excessive concentration of nitrate in the water intake may result from other causes than the low level of the water table. It can result from weak security of the well [29]. There have been cases of wastewater introduction through poorly protected old wells or open piezometers, directly into the aquifer. Such action poses a serious risk to the quality of groundwater. In addition, as well as the increase in ammonia concentration, other water quality indicators (e.g. microbiological indicators) deteriorate.

One of the most dangerous sources of water contamination by nitrogen compounds is the manner of storage of livestock manure. In Poland, this method of manure management is often practiced. This is due to insufficient awareness and ignorance of farmers. Investment in manure slabs and liquid manure tanks does not solve the problem. Proper manure management should be consistent with the principles of good farming practice [30]. It has been found that long-term storage of manure (on the ground, in a pile) results in severe soil and groundwater pollution with nitrogen compounds, in particular nitrogen and ammonium nitrate. At the same time, short-term (half-yearly) storage of manure in the pile did not pose a serious threat to water quality. For these reasons, with a view to the protection of water quality, livestock manure should not be stored directly on the ground for extended periods (short-term collection is considered acceptable, but verification needs to be made on the basis of additional testing) [31].

The impact of anthropogenicity on groundwater pollution can also be significant away from farms. Water samples, taken in such places, may confirm (and often do confirm) the assumption of local impact of such pollutants. The lack of proper water and sewage management is a decisive factor in water pollution [32]. The problem of groundwater quality is often noted in the vicinity of settlement units, where the management of domestic sewage mainly takes place through a sewage holding tank system. The relationship between wastewater management and the quality of groundwater can be unequivocally demonstrated. It is confirmed by the presence of indicator bacteria in waters, adjacent to leaky tanks and uncontrolled discharges of wastewater into the environment [33]. It is also possible to use isotopic and geochemical properties, especially with respect to nitrates (NO_3^-) [34]. In Poland, the scale of uncontrolled discharge of wastewater into the environment is huge. The problem of the impact of such activities on the groundwater quality has been widely discussed in post-mentioned work [35].

Due to the problems associated with the use of groundwater resources in rural areas, the management of these resources should be integrated into regional spatial planning. The aim should be to achieve a balance between ecological and engineering approaches [36]. Among others, the changes in water quality should be continuously monitored to ensure optimum use of sources. When it is justified and possible, the exploited source should be replaced by sources of better quality [37]. It is also recommended that groundwater intake in rural areas be related to levels of aquifer replenishment.

3 The quality of groundwater in the aspect of the functioning of selected rural water intakes

The small water intakes presented in this work are typical rural systems, based on the groundwater. In the first decade of the 21st century, the quality of the groundwater was high enough that, in the described intakes, water did not require pro-quality activities. Water was directly channeled to the customers. In other places, water was effectively treated in water treatment plants, by aeration, subsequent filtration and final disinfection. However, in recent years there has been a significant deterioration in the quality of the intake water. Lower water quality and limited local water resources have resulted in difficulties in supplying residents with drinking water of the required quality [38]. These issues are presented in the following part of the work.

Delivery of drinking water to the inhabitants of A village (137 people) takes place through a groundwater intake, consisting of one well at a depth of 39.0 m below ground level. The well has been in operation since 1981. The aquifer is in the form of gravel, where the water flows under pressure. The aquifer is located within Quaternary deposits. The quality problems of this water are related, among others, to the content of ammonium ions in the intake water. These ions are present in the water in the form of N-NH4. The concentration of ammonium ions in water has increased several times in recent years. In 2016 it reached a value of 2700 μg/l (Fig. 1) [39].

In the authors' view, the causes of the presented qualitative changes in water may be the result of transformation of the local hydrogeochemical environment and the changes in the water circulation system caused by the mixing of waters from various hydrogeochemical environments.

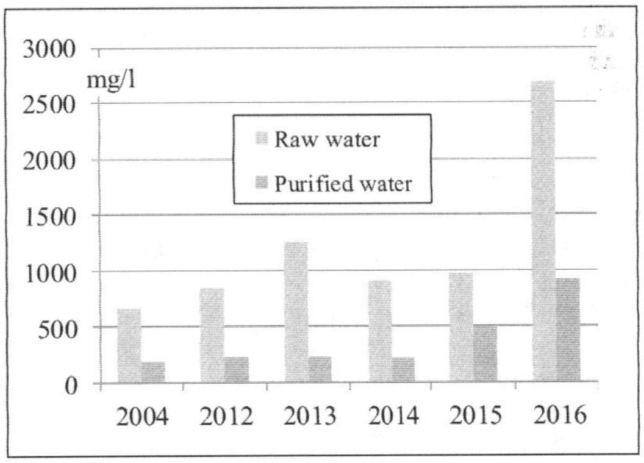

Fig. 1. The concentration of ammonium ions in the raw and treated water in A, the year 2004 and the years 2012-2016.

The above thesis is confirmed by the results of the investigations into iron and manganese concentrations (described below) which, as a result of described changes in the water-soil environment (i.e. with ammonium ion), were also increased [13]. It is also important that the water is only taken through one well, which drains the groundwater resources only in one place. The local water company decided to use ion exchange columns to obtain the required water treatment parameters. The authors consider the adopted solution to be appropriate, taking into account technical and economical aspects. Removing the ammonium ions in the process of intensive aeration of water would be ineffective in the given situation (the ammonium ion is in the form of N-NH4). In addition, high concentrations of N-NH4 would make it difficult to remove them during the process of water filtration, i.e. during iron and manganese removal.

In A village, in addition to the problems of raw water quality presented, increased concentrations of iron and manganese (to a lesser degree) have been found in the water in recent years. The most likely cause of the changes has been previously presented by the authors of the work. Within the period of ten years (from 2004 to 2014) the concentration of iron in water increased several times, reaching a level of 4500 mg/l Fe in 2016. This increase had a significant impact on the effects of water treatment (Fig. 2) [39].

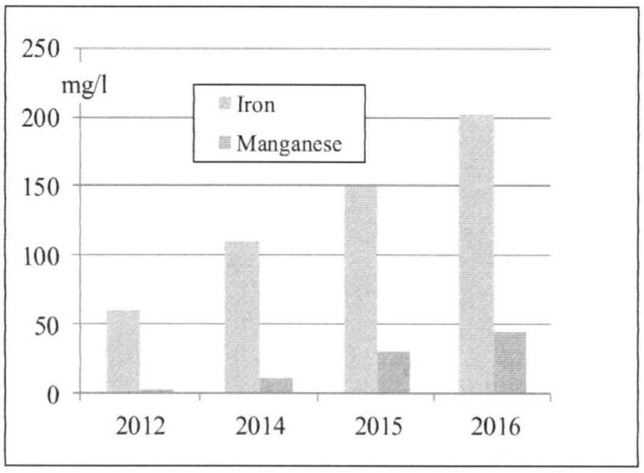

Fig. 2. The concentration of iron and manganese in purified water in A; the year 2012 and the years 2012-2016.

Modifications in the water treatment station included a reduction in filtration rate, frequent washing of the filter beds as a result of clogging of the filters and the necessity to frequently clean the aerators or replacement of them. In addition, to improve the aeration process in the air/water mixer, the amount of oxygen supplied was increased.

A different qualitative problem, concerning the intake water, was found in B village. The local water intake supplies water to three villages, inhabited by a total of 524 inhabitants. Until recently, the intake water to customers was supplied without treatment, because the water quality met the appropriate quality requirements. It should be added that water is taken from one well (43 m deep below ground level which has been in operation since 1975). The concentration of nitrate in the intake water increased gradually from 1989 to 2016 (Fig. 3) [39].

The cause of the qualitative changes in water is, according to the authors, the agricultural activity around the location of the water intake (cereal and strawberry cultivation, fertilization). In addition, long-term storage of manure took place near the water intake. An ion exchange process was used to improve the water quality: the owner installed two ion exchange columns, and regenerated the bed with brine. However, the situation should not be considered to be controlled. Since 2014, besides elevated nitrate concentrations, coliforms and escherichia coli appear sporadically in the water being tested. The effect is to close the intake for the time taken to disinfect the water. During the shutdown period, water is delivered to the customers in barrels. The presence of bacteria is related to poorly managed sewage management in the nearest rural areas [35].

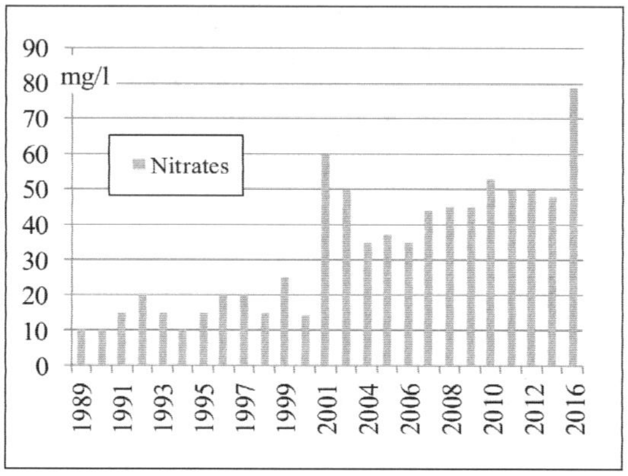

Fig. 3. The concentration of nitrates in raw water in B; the years 1989÷2016.

The water from the intake in C village is supplied to two villages with a total population of 133 inhabitants. Commencing in 2016, excessive concentrations of manganese have been observed in the water and, in addition, microbial contamination is increasingly occurring in the water (this problem has been observed since March 2015). There are two wells - the first well with a depth of 28 m below ground level, currently working as a back-up (drilled in 1966) and the second well with a depth of 24 m below ground level (in operation since 1969). From October 2016, the total number of bacteria in the treated water is constantly above 100 cfu/1 ml despite cyclical disinfection of the wells. There are also coliforms present in amounts of 2÷3 cfu/100 ml to 15 cfu/100 ml. The bacteriological hazard of water was eliminated by continuous water disinfection with chlorine. According to a geological study [40], neither well in C village has natural insulation. The water intake includes the upper interglacial level, which is in the region of the uncovered level. In this case, the location of the intake and its geological characteristics are extremely important for water safety, particularly from a microbiological point of view. This is due to inadequate sewage management in the nearest villages. In the vicinity of this water intake (the nearest villages), the scale of leakage of sewage collection and treatment system is over 90% [35]. For these reasons, the authors stress the important role of protection zones, both direct and indirect. In the case of the described water intake, these zones could reduce the negative impact of wastewater on the quality of groundwater.

The water intake in D village supplies the water to the 141 residents of this village. The groundwater is taken from a depth of 37 m below ground level through a drilled well. The well has been operational since 1969. The water is directed to a water treatment plant where it is treated in a conventional system. However, the increase in the concentration of iron in the intake water (from 2500 µg Fe/l in 2012 to 5110 µg Fe/l in 2016) has caused operational problems at the WTP. The deterioration in the quality of the water is due to changes in the local hydrogeochemical environment. It is also the effect of the intake of water from one well. The consequence of these phenomena was the deterioration in the efficiency of water treatment. The concentration of manganese in treated water increased. Within the pro-quality activities, the pressure filter was replaced, and an additional water-air mixer was installed to increase the effectiveness of manganese removal from water. It was assumed that the problems with manganese were due to an excessively low concentration of oxygen in the water. The amount of oxygen in the water after the iron removal process was insufficient to oxidize the manganese. In order to remove iron from the water efficiently (despite the continuous increase of its concentration in raw water), the velocity of water flow through the filter bed has been reduced. At the same time, a high frequency of degasser cleaning was adopted. Large amounts of precipitated iron were deposited on elements of equipment and pipes (Fig. 4).

Fig. 4. The part of the installation (elbow) from the water treatment plant in D.

In E village, there is a water intake, supplying 162 residents of the village with water for consumption. The water intake is based on one well with a depth 116.5 m below ground level. The well has been in operation since 1976. Up to 2006, the water was directed to the recipients without treatment. In that year, in order to provide customers with water of a suitable quality, a water treatment system was installed. The system was based on aeration and water filtration in pressurized equipment. Silica sand and G-1 catalytic mass were used to fill the filter bed. By 2014, the iron concentration in the water was at a more or less constant level. It ranged from 300 to 400 µg Fe/l. However, in 2015 the iron concentration increased to 745 µg Fe/l. In the next year, an increase in the manganese concentration in intake water was found (434 µg/l Mn). During the previous three years the average manganese concentration was 200 µg/l Mn. (It varied between 150 and 220 µg/l Mn) It is the opinion of the authors that the significant increase in iron and manganese concentrations in water is due to similar reasons to those presented in A and D villages.

The summary table (Tabela 1) of checked water quality and changes, for all described water intakes, it was placed below.

Table 1. The changes in water quality - a summary.

A village		
The concentration (µg/l) in year:	2004	2016
ammonium in raw water	660	2700
ammonium in treated water	180	920
The concentration (µg/l) in year:	2004	2016
iron in raw water	1100	4500
The concentration (µg/l) in year:	2012	2016
iron in treated water	60	203
manganese in treated water	2	45
B village		
The concentration (mg/l) in year:	1989	2016
nitrate in raw water	10	73
Bacteria (cfu/100 ml) in year:	2014	
escherichia coli, coliforms	sporadically	
C village		
Bacteria (cfu/100 ml) in year:	2016	
the total number of bacteria	to 100	
coliforms	to 15	
D village		
The concentration (µg/l) in year:	2012	2016
iron in raw water	2500	5110
E village		
The concentration (µg/l) in year:	to 2014	2015
iron in raw water	300-400	745
The concentration (µg/l) in year:	to 2015	2016
manganese in raw water	150-220	434

4 Conclusions

- The quality of groundwater in rural areas is strongly dependent on anthropogenic factors, particularly the type and intensity of agricultural activities and the method of wastewater management. For example, in B village a high nitrate concentration in water was found. The concentration of these compounds in the intake water increased gradually from 1989 (10 mg N-NO3/l) to 2016 (73 mg N-NO3/l). Since 2014, besides elevated nitrate concentrations, coliforms and escherichia coli appear sporadically in the water being tested. The cause of the qualitative changes in water is, according to the authors, the agricultural activity (cereal and strawberry cultivation, fertilization) and long-term storage of manure around the location of the water intake.
- The major water quality problem in small water supply

systems is the high concentrations of iron, manganese, nitrogen compounds (nitrates and ammonium) and bacteriological contamination. For example, in E village the iron concentration in the water was at a more or less constant level by 2014 (it ranged from 300 to 400 μg Fe/l). However, in 2015 the iron concentration increased to 745 μg Fe/l. In the next year, an increase in the manganese concentration in intake water was found (434 μg/l Mn). During the previous three years the average manganese concentration was 200 μg/l Mn.

- In rural water supply systems, protection zones are required, both direct and indirect. The zones could reduce the degradation of groundwater resources (especially shallow groundwater).
- Continuous monitoring of the quality of the water taken is required. This is very important for small water intakes, especially where one well is used. This is related to a decrease in the quality of intake water, which negatively affects the efficiency of its treatment systems. For example, in D village the increase in iron concentration in the intake water (from 2500 μg Fe/l in 2012 to 5110 μg Fe/l in 2016) has caused operational problems at the WTP. In particular, an increase in the concentration of manganese in treated water was observed. In D village, the deterioration in the quality of the water is also the effect of the intake of water from one well.
- It is recommended for water intakes equipped with one well, that they be supplemented with more wells. Additional wells can help with the work of the intake of water (e.g. alternate pump work, security in case of failure and reduce overexploitation of the aquifer).

References

1. E. Radzka, J. Jankowska, Adm. Zarządz. **107**, (2015)
2. GUS. Warszawa. *Ochr. Środ. 2016,* (2017)
3. Z. Michalczyk, Woda Środ. Obsz. Wiejs. **4**, 11 (2004)
4. R. Nowak, A. Imperowicz, Instal **364,** 7/8 (2015)
5. D.N. Lerner, B. Harris, LUP **26,** 1 (2009)
6. W. Mioduszewski, Woda Środ. Obsz. Wiejs. **6**, 1 (2006)
7. S.S.D. Foster, P. J. Chilton, *Philosophical Transaction B of The Royal Society B* (2003)
8. M. Guzik, Biuletyn *PIG* **432**, (2008)
9. E. Muntean, T. Mihaiescu, Environm. Eng. Manag. J. **15,** 12 (2016)
10. J. Raczuk, K. Sarnowska, Arch. Ochr. Środ. **28**, 3 (2002)
11. L.L. Gan, CHIM Oggi **34**, 6B (2016)
12. M. Gutry-Korycka, A. Sadurski, Z.W. Kundzewicz, J. Pociask-Karteczka, L. Skrzypczyk, *Nauka* **1**, (2014)
13. J. Górski, *Zaopatrz. w wodę jakość i ochr. wód: zag. współcz.* (ed. M.M. Sozański, Poznań, 2010)
14. A. Sapek, Woda Środ. Obsz. Wiejs. **8,** 1 (2008)
15. E. Janigacz, Prz. Geolog. **47,** 3 (1999)
16. I. Kajewski, Infrastr. Ekol. Ter. Wiejs. **8**, 1 (2010)
17. B. Sapek, Woda Środ. Obsz. Wiejs. **6**, 1 (2006)
18. J. Jaszczyński, A. Sapek, S. Chrzanowski, Woda Środ. Obsz. Wiejs. **6**, 2 (2006)
19. S. Zabłocki, Biuletyn PIG **456**, 2 (2013)
20. J. Raczuk, E. Biardzka, M. Michalczyk, Woda Środ. Obsz. Wiejs. **9**, 1 (2009)
21. I. Tałałaj, Infrastr. Ekol. Ter. Wiejs. **5**, (2008)
22. M. Fic, W. Mioduszewski, Woda Środ. Obsz. Wiejs. **3**, 6 (2003)
23. M. Guzik, Przeg. Geolog. **51**, 2 (2003)
24. M. Basu, S. Hoshino, S. Hashimoto, Nat. Hazard. Rev. **81**, 1 (2016)
25. A.M. Marciniak, P. Wesolowski, A. Brysiewicz, P. Burczyk, J. Elem. **21**, 2 (2016)
26. T.Y. Stigter, L. Ribeiro, A.M.M. Carvalho Dill, J. Hydrol. **327**, 3/4 (2006)
27. J.W. Chen, H. Qian, Exposure Health **8**, 3 (2016)
28. K. Pawęska, B. Malczewska, A. Bawiec, A. Bauerek, JEE **49**, (2016)
29. K. Pawęska, B. Malczewska, B. Zyglińska, *Proceedings of ECOpole* **6**, 1 (2012)
30. T.K. Woroniecki, E. Rumasz-Rudnicka, Acta Agroph. **11**, 2 (2008)
31. S. Pietrzak, P. Nawalany, Woda Środ. Obsz. Wiejs. **8**, 2b (2008)
32. J. Kachnic, Prz. Geolog. **49**, 2 (2001)
33. B.G. Katz, D.W. Griffin, P.B. McMahon, H.S. Harden, E. Wade, R.W. Hicks, J.P. Chanton, J. Environ. Qual. **39**, 4 (2009)
34. I. Zimoch, J. Paciej, Desalin. Water Treat. **57**, 3 (2016)
35. R. Nowak, A. Imperowicz, Inż. Ekol. **47** (2016)
36. M.L. Collin, A.J. Melloul, J. Cleaner Prod. **11**, 7 (2003)
37. Z. Yousefi, S. Rouhani, S.M. Ansari Yeganeh, Iranian J. Health Sc. **2**, 1 (2014)
38. Regulation of the Minister of Health of 13 November 2015 on the quality of water intended for human consumption, J. Laws, item 1989 (2015)
39. Own study based on data from local water companies
40. Opinion to determine the hydrogeological conditions in the area of groundwater intakes in C village

The occurrence of priority substances in surface water

Katarzyna Moraczewska-Majkut[1,*], and *Katarzyna* Kalemba[2]

[1] The Silesian University of Technology, Faculty of Energy and Environmental Engineering, Institute of Water and Wastewater Engineering, Konarskiego 18, 44-100, Gliwice, Poland
[2] The Silesian University of Technology, Faculty of Energy and Environmental Engineering, Institute of Water and Wastewater Engineering, Konarskiego 18, 44-100, Gliwice, Poland

Abstract. In the article the presence of substances, specified in the Water Framework Directive as a priority, in the aquatic environment was analyzed. The impact of priority substances on organisms - especially on human health and ways of reducing their occurrence was discussed. Particular attention was paid to the legal aspects of the presence of these compounds in the aquatic environment and the principles of monitoring of priority substances.

1 Introduction

The substances known as toxic are compounds that cause life-threatening organisms or their death. Toxic contaminations have a major impact on the quality of surface water. Concentrations of substances in excess of environmental quality standards can cause poisoning and death of aquatic organisms. The awareness of this threat has led to the emergence of legislation in the European Union that aims to protect the environment and, in particular, protect water from chemical pollution. Mainly, in the aquatic environment pollutants migrate between particular elements of the environment [1].

In Directive 2000/60/EC of the European Parliament and of the Council of 23 October 2000 establishing a framework for Community action in the field of water policy known as the Water Framework Directive, 33 priority substances were identified, 13 of which were identified as priority hazardous substances. These concepts are defined in the above directive: "Prioritized substances are substances that pose a particular threat to the aquatic environment," while "priority substances are substances or groups of substances that pose significant environmental risks due to their persistence, toxicity, bioaccumulation, low susceptibility to degradation and risk to human health "[1]. Currently, 45 substances are in the list of priority substances in the field of water policy [2].

2 Analysis of the presence of priority substances in surface waters

The world wide pollution of freshwater ecosystems with thousands of industrial chemicals is increasing, which is crucial problem of modern environmental protection. More than a third of the available fresh water on Earth is used for agricultural, industrial and domestic purposes and all these activities lead to the contamination of waters with numerous synthetic compounds. Every year, many hazardous chemicals use in industrial and consumer products can in part get into natural waters. Additional, pollution come from diffuse sources, including from agriculture, where millions of tons of pesticides are used each year [3].

Understanding the causes and sources of water pollution by priority substances allows Member States of the European Union to better control these pollutants. The main sources of pollutants in these waters are atmospheric deposition, surface runoff, sewage discharges and leachate from landfills [4]. Pollution present in the atmosphere can settle on the entire surface of the catchment and can be washed away by atmospheric precipitation, becoming part of surface runoff of rainwater. Pollutants can also penetrate directly from atmospheres into water bodies, rivers or coastal waters. Substances can be present in the atmosphere either as gaseous particles, solid particles or liquid. All these compounds settle on the surface of the Earth through wet or dry deposits [5].

Urban pollution has a significant impact on the quality of all ecosystems. In particular, changes in the use of land in urban planning areas significantly change the condition of the water environment, inhibit the seeping of rainwater into the ground and cause flooding. In addition, urbanization has a strong influence on the quality of rainwater, which is increasingly contaminated by biological and physicochemical substances of anthropogenic origin. Contaminated rainwater is one of the major sources of pollutants released into waters [6,7].

Studies have shown that waste water treatment plants are a significant source of water pollution by priority substances. The waste water come from industrial plants or urban areas contain pollutants such as polycyclic aromatic hydrocarbons, phthalates, phenols, phenols, flame retardants and others. The fact that waste water treatment plants are designed primarily to remove organic matter, nitrogen and phosphorus from sewage

* Corresponding author: katarzyna.moraczewska-majkut@polsl.pl

and there are no legal regulations on the concentration of priority substances in sewage causes many of these substances to pass from conventional treatment plants to surface water, causing them to contaminate [8,9]. Landfills are also a source of environmental pollution by priority substances. The big problem is that these compounds do not appear as separate fractions but as additives to stored products such as packaging, plastics, textiles or electronics. In addition, leachate effluents contain high concentrations of suspended and dissolved organic matter, which increases the potential for transport of these contaminants [10-13].

3 Effect of priority substances on living organisms

Priority substances are substances that are particularly harmful to the health and life of organisms. Presence in the waters of these substances and in particular concentrations above acceptable levels, can cause poisoning as well as extinction not only of aquatic organisms, but also of benefit or contact with human and animal water [14-17]. High toxicity, persistence, and the ability of priority substances to bioaccumulate are the most dangerous factors for health and life. Much of the low water-soluble priority substances are readily soluble in fatty tissues of living organisms. This causes the concentration of these toxins in human and animal organisms much higher than in the environment. This process, called bioaccumulation, means that organisms at the end of the food pyramid and above all people are exposed to the accumulation of toxic substances in their bodies [18-25]. Heavy metals, such as lead, cadmium and mercury in the priority substance group, interfere with the function of living organisms by destroying or impairing their cells. Cadmium causes permanent damage to the kidneys and the skeletal system, lung cancer and heart disease [25,26]. Mercury damages the nervous system [19]. Lead is responsible for anemia and for impairment of the function of the nervous system [27,28]. Scientific studies also prove that many of the substances in question cause cancer, disrupt the effects of the nervous and hormonal systems of humans and animals, cause foetus damage, mutations and damage to internal organs, especially liver [14-28].

4 Surface water monitoring for the presence of priority substances

Within the framework of the State Environmental Monitoring, the quality of surface waters is assessed, which results from the Water Law Act [29]. Its purpose is to provide a basis for action to improve water quality and protect against pollution from the agro-industrial, agricultural and industrial sectors - including protection against salinity and priority substances [30].

Water quality measurements are carried out at the measuring and control points of the rivers and in the transitional and coastal waters of the Baltic Sea, by the Voivodship Inspectorates for Environmental Protection [30]. The assessment of the state of the water is based on the classification according to the Regulation of the Minister of the Environment of 21 July 2016 on the classification of the state of surface water bodies and environmental quality standards for priority substances [31]. The state of unitary bodies of surface water is classified on the basis of the results of the studies conducted for the presence of priority substances and other pollutants in these waters. The classification of the chemical state of the waters is based on at least 12 measurement results from the monitoring of priority substances comparing them to the environmental quality standards for each category of surface water (contained in Annex 9 of the Regulation). Classification of individual physico-chemical elements takes place annually [31].

According to the Water Law [29], the monitoring consists of studies on the tissues of living organisms of five priority substances: diclofen, perfluorooctane-sulfonic acid and its derivatives (PFOS), dioxins and dioxin-like compounds, hexabromocyclo-dodecane (HBCDD) and heptachlor and heptachlor epoxide. It belongs to the Chief Inspectorate of Environmental Protection. From 2019 laboratory analyzes for seven of the "new" substances (quinoxyfen, aclonifen, bifenox, cyberatine, cypermethrin, dichlorophos and terbutryne) will be carried out by four provincial environmental inspectorates. In 2015, the Chief Inspector of Environmental Protection has appointed the Provincial Inspectorate for Environmental Protection for this purpose: in Lublin, Szczecin, Łódź and Bialystok.

4.1 Ranking of priority substances based on water monitoring data

Ranking of priority substances is based on data from the monitoring carried out by the Voivodship Inspectorates for Environmental Protection in order to identify those priority substances whose concentration exceeds the limits and then determine the validity of these pollutants. When compiling the ranking of priority substances, their ecological sustainability, toxicity, frequency of exceeds and environmental discharges are taken into account. Ranking is prepared in three stages. During the first stage, the frequency of exceeding the average annual values of water quality standards in relation to the number of determinations of a given pollutant at each control and measurement point of the water body is determined. Using these data, priority substances are ranked in descending order. The second step is to determine the frequency of the priority substance in relation to the number of measurements performed at all control points. The concentration of the substance is greater than the detection threshold but lower than the concentration limit. Just like before, the results are laid down in a descending order. In the third step, the rest of the substance that was not classified during the first or second stage is placed on the previously prepared ranking list of priority substances from the first and second stage [32].

4.2 Ranking of priority substances based on emission data

The ranking of priority substances can also be made according to the assessment of the priority substances released into the environment. Priority substances are ranked each year in descending order according to the total discharged volume in a given region [32]. Data on discharges can be found in the European Pollutant Release and Transfer Register (E-PRTR) [32]. Regulation (EC) No 166/2006 of the European Parliament and of the Council of 18 January 2006 on the establishment of a European Pollutant Release and Transfer Register and amending Council Directive 91/689/EEC and 96/61/EC established an integrated European Pollutant Release and Transfer Register Pollution (E-PRTR), which is commonly available in the electronic database [33]. Each EU Member State was obliged to establish a National Register and Transfer of Pollution (PRTR) [34]. Poland fulfills this requirement by the Ordinance of the Minister of the Environment of 14 August 2009 on the report for the creation of the National Register of Discharges and Transfer of Wastes (No 141, item 1515), which specifies the specimen form for the report to be made National Pollutant Release and Transfer Register, form and required techniques for submitting such report [33]. PRTR collects data on pesticides, heavy metals, non-organic substances, organic chlorinated substances and other information including priority substances. The register contains data on releases of these pollutants into water, soil, air, their off-site transfers, pollutants in sewage, releases of pollutants from diffuse sources. This information can be searched in the database according to a number of criteria such as the type of pollutants, the geographical location of the industrial plant, the environmental elements to which the pollutant is released, according to the plant owner or the type of industry [34].

Types of activities requiring PRTR reports are listed in Annex I of the Regulation and are [33,34]: energy industry (e.g. oil and gas refineries, electro-flare plants), metal production and processing (e.g. ore roasting and sintering installations chemical industry (e.g. chemical installations for the production of basic organic and inorganic compounds, fertilizers, biocides, plant protection products), waste management and waste management, waste water (e.g. waste water treatment plants, waste landfills), paper and wood production and processing, intensive poultry or pig farming, other activities such as tanneries or dairies. An operator of at least one establishment engaged in the activities listed in Annex I to this Regulation is obliged to report annually to the competent national authority on the quantitative releases of pollutants in Annex II of that Regulation (including all priority substances) to air, water and off-site transfers. These data can be based on measurement, calculation or estimation and are reported in unit kg/year and waste in unit Mg/year. Member States shall set a deadline by which all facility operators communicate the data to the authorities. The data is then transmitted electronically to the European Commission [34].

5 Conclusions

Priority substances that occur in surface or underground water can be natural and anthropogenic origin and belong to many different groups of substances such as hydrocarbons, heavy metals, organometallic compounds, chloroorganic compounds, polycyclic aromatic hydrocarbons and others. A large proportion of all priority substances are pesticides. The vast majority of priority substances exhibit toxicity to living organisms. The Water Framework requires the presence of contaminants in the waters to be monitored. Data collected during the monitoring of waters are used, among others, to classify the state of surface water bodies or to create rankings of priority substances. Ultimately, the aim is to bring the environment to a state in which the concentration of priority substances will not exceed the intrinsic background concentration of the aquatic ecosystem.

Determining the presence in surface waters of priority substances and identifying sources of water pollution allows action to be taken to improve water quality. Good water status can be achieved mainly by eliminating and reducing discharges of these substances by industrial plants or waste water treatment plants.

References

1. Directive 2000/60/EC of the European Parliament and of the Council of 23 October 2000 establishing a framework for Community action in the field of water policy.
2. Ordinance of the Minister of the Environment of 6 May 2016 on the list of priority substances, Journal of Laws from 2016 item 681.
3. Dichlorvos. EXTOXNET 1996. http://extoxnet.orst.edu/pips/dichlorv.htm, data of access: 15.04.2017.
4. Diuron. Chemical Fact Sheet. http://www.speclab.com/compund, Spectrum Laboratories, Inc., data of access: 17.04.2017.
5. A. Bressy, M.C. Gromaire, C. Lorgeoux, M. Saad, F. Leroy, G. Chebbo, Water Res. **46** (2012)
6. R. Pitt, R. Field, M. Lalor, M. Brown, Water Environ. Res. **67** (1995)
7. S. Zgheib, R. Moilleron, M. Saad, G. Chebbo, Water Res. **46** (2012)
8. N. Marti, D. Aguado, L. Segovia-Martinez, A. Bouzas, A. Seco Mar. Pollut. Bull. **62** (2011)
9. J. Robles-Molina, B. Gilbert-Lopez, J.F. Garcia Reyes, A. Molina-Diaz, Talanta. **82** (2010)
10. L. Patrolecco, N. Ademollo, S. Capri, R. Pagnotta, S. Polesello, Chemosphere **81**, 11, (2010)
11. M. Gono, M. Kyncl, R. Gono, I. Kłosok-Bazan, P E **11**, 89 (2013)
12. I. Zimoch, B. Kotlarczyk, A. Sołtysik Ochr. Sr. **29**, 3 (2007)

http://www.os.not.pl/docs/czasopismo/2007/Zimoch_3-2007.pdf

13. J. Boguniewicz-Zablocka, I. Klosok-Bazan, V.Naddeo, Environ Sci Pollut Res (2017). https://doi.org/10.1007/s11356-017-0608-8
14. G. Svecevicius, *Acute Toxicity of Nickel to Five Species of Freshwater Fish* (Institute of Ecology of Vilnius University. Vilnius, 2009)
15. A. Tuvikene, *Responses of fish to polycyclic aromatic hydrocarbons (PAHs)* (Finnish Zoological and Botanical Publishing Board. Helsinki, 1995)
16. Environmental health criteria. Lead – environmental aspects. World Health Organization. Geneva, 1989
17. European Union Risk Assessment Report. Alkanes, C10-13, chloro. Institute for Health and Consumer Protection Office for Official Publications of the European Communities. Italy, 2000.
18. R. Eisler, Polycyclic Aromatic Hydrocarbon Hazards to Fish, wildlife and Invertebrates: Synoptic Review. U.S. Fish and Wildlife Service, 1987.
19. J.F. Risher, Elemnetal Merkury and inorganic Merkury compounds: human health aspects, World Health Organization, 2003.
20. Environmental Quality Standards (EQS) Substance Data Sheet. Common Implementation Strategy for the Water Framework Directive. Brussels, 2005.
21. European Union Risk Assessment Report. Alkanes, C10-13, chloro. Institute for Health and Consumer Protection Office for Official Publications of the European Communities. Italy, 2000.
22. I .Zimoch, E. Łobos, Desalin. and Water Treat. **52**, (2014) https://doi.org/10.1080/19443994.2014.884684
23. I. Zimoch, Ochr. Sr. **29**, 4 (2007) http://www.os.not.pl/docs/czasopismo/2007/Zimoch_4-2007.pdf
24. M. Bożym, I. Kłosok-Bazan, M. Wzorek, Pol. J. Environ. Stud. **27**, 3 (2018) DOI: 10.15244/pjoes/75828
25. Cadmium In Drinking-water. Background document for development of WHO Guidelines for Drinking-water Quality, World Health Organization, 2011.
26. Toxicological profile for cadmium. Agency for Toxic Substances and Disease Registry (ATSDR), 2012.
27. Agency for Toxic Substances and Disease Registry, Toxicological profil for lead, U.S. Dprt. of Health and Human Services, 2007.
28. *Environmental Handbook Volume III: Compendium of Environmental Standards* (GTZ; 1995)
29. Act of 18 July 2001 Water Law, Journal of Laws from 2001 No. 115 item 1229.
30. Dangerous substances for the Baltic Sea environment - properties, sources and reduction of emissions. COHIBA. Institute of Ecology of Industrial Areas. 2012.
31. Regulation (EC) No 1907/2006 of the European Parliament and of the Council of 18 December 2006 (REACH).
32. Directive 2010/75/EU of the European Parliament and of the Council of 24 November 2010 on industrial emissions (Integrated Pollution Prevention and Control).
33. F. Oosterhuis, Substitution of hazardous chemicals. Institute for Environmental Studies (IVM). Vrije Universiteit. Amsterdam, 2006.
34. http://www.katowice.pios.gov.pl/monitoring/pms/wody.pdf, data of access: 08.04.2017.

The reduction of 2- and 3-ring PAHs entering to the surface waters in the integrated processes

Maria Włodarczyk-Makuła[1,*], and *Agnieszka* Popenda[1]

[1]Czestochowa University of Technology, 42-200 Czestochowa, Poland

Abstract. The aim of the study was to determine the removal of 2- and 3-ring polycyclic aromatic hydrocarbons (PAHs) during oxidation process and during simultaneous oxidation and adsorption processes. Technological research was conducted with the use of biologically treated industrial wastewater. Oxidation was carried out with dihydrogen peroxide. The adsorption process was carried out onto activated carbon. Quantitative and qualitative determination of polycyclic aromatic hydrocarbons was conducted using gas chromatograph and mass spectrometer system. The experiments were conducted under constant pH as well asunder the constant temperature. The oxidation efficiency of the analyzed hydrocarbons under oxidation conditions ranged from 14 to 81%. During the simultaneous oxidation process and adsorption on activated carbon, the PAHs removal ranged from 43 to 100%. The results lead to the conclusions that simultaneous oxidation and adsorption can be successfully apply in the final treatment of industrial wastewater resulting in the minimizing PAHs load entering to the surface waters.

1 Introduction

Recently, in technology wastewater most often the third step of wastewater treating by applying additional processes is of high importance. The attention is paid into increasing the level of removal both organic and inorganic pollutants resulting in lowering the load of pollutants entering to the surface waters. That is important because surface waters are the sources of water in the municipal administration systems. In current legislations the presence of organic micropollutants, such as: polycyclic aromatic hydrocarbons (PAHs) is omitted [1]. On the other hand PAHs are listed among contaminants that should be removed from wastewater due to their carcinogenic and mutagenic properties. So far, the investigations into treated wastewater have indicated that although the high level of wastewater treatment is obtained, the efficiency removal of PAHs is low [2,3]. This phenomenon was observed both in municipal and industrial wastewater. With respect to industrial wastewater the certain load of micropollutants is removed in the industrial factories however, more strict requirements of environmental quality force the application of methods and processes in the removal or decomposition of entering to the receiver pollutants. It is mainly aimed to organic pollutants, such as: polycyclic aromatic hydrocarbons (PAHs), organic chlorine derivatives or colouring agents [4,5]. Some of them have carcinogenic and mutagenic influence on the organisms as well as they are relative persistent in the environment.

The literature studies and own investigations indicate that is possible to apply sorption as well as advanced oxidation processes (AOP) with respect to PAHs removal [6-9]. Considering sorption process the activated carbons, including these obtained from various waste materials, are mostly applied in the removal of PAHs. Among wide range of the literary sources regarding sorption of organic substances onto activated carbons the limited information can be found of the applying sorption in the removal of PAHs from wastewater [6,10]. The available literature [11-15] confirm that activated carbons may be successfully apply in the removal of organic compounds from the water solutions. It should also be highlighted that in the water solutions, such as: waters, wastewater, soil-water environment the mixture of various organic compounds appears resulting in the limited efficiency of sorption of selected compounds due to the possibility of the presence of competitive sorption and suspensions. Therefore, it is searching for alternative methods of removal of organic compounds from the water solutions including oxidation in the advanced oxidation processes [16,17]. The literary data [9,18-20] confirm the high efficiency of AOP in the removal of organic substances especially resistant to the biodegradation. The fundamentals of the afore mentioned methods is oxidation of most organic compounds to the CO_2, H_2O with the usage of radicals hydroxyl OH•. Dihydrogen dioxide or a compound of Fenton's reagent is one of the oxidants that can be apply in the AOP processes [21-23]. The application of Fenton's reaction in the oxidation of selected PAHs was carried out in the studies of Rubio-Clemente [24]. So far, the investigations aiming the advanced oxidation of PAHs have been conducted on model solutions containing individual or no more than few hydrocarbons. It should be highlighted that PAHs

* Corresponding author: mwm@is.pcz.czest.pl

are present in the environment as mixtures of various composition. In the literary there is also information available of the studies of the removal organic compounds in two-step process - oxidation and sorption. It is described the primary removal of organic compounds in the sorption, and then oxidation of adsorbed substances. Simultaneous sorption and oxidation can be regarded as another system of integration the afore mentioned processes. In this case adsorbent (e.g. activated carbon) plays the role of sorbent of organic compounds as well as catalyst of radicals hydroxyl formation responsible for the oxidation of organic substances: dissolved and adsorbed compounds [17,23]. Although many investigations into advanced oxidation processes and adsorption have been carried out up, the decomposition of PAHs in the industrial wastewater under oxidation in the presence of activated carbon has not been considered so far. Due to the satisfactory effects of the removal/degradation of PAHs both during the sorption and oxidation the studies aiming the determination of removal efficiency of hydrocarbons by combing both process have been undertaken [3,4,23,24].The aim of the work was to determine the removal efficiency of 2- and 3-ring PAHs during simultaneously conducted oxidation with dihydrogen dioxide and sorption on activated carbon.

2 Methodology of technological studies

2.1 Material and methodology of conducting investigations

The technological studies were carried out using biologically treated coke-plant wastewater. The wastewater samples were taken once. In raw wastewater samples the initial organic content was determined by chemical oxygen demand (COD) as well as PAHs. It should be noted that coke wastewater originating from the technological processes enter to the biological wastewater treatment plant working with activated sludge and the nitrification of ammonia nitrogen removal. The technological studies were carried out with the usage of strong oxidant dihydrogen dioxide (30%). The dose of the oxidant equal to 10ml/L was formerly adjusted [24,25]. The sorption was carried out using activated carbon. The time of carrying the simultaneous process of oxidation and adsorption was constant and equal to 1 hour. Then, the concentration of selected hydrocarbons and organic content expressed as COD was determined.

2.2 Analytical methods of PAHs

In order to determine PAHs, the initially preparation of wastewater samples was performed during extraction of organic substances by liquid-liquid method. Mixture of cyclohexane and dichlomethane (v/v – 5:1) was added to the wastewater samples. Samples containing activated carbon were filtered and then both solution and suspension samples were analyzed. In this case to the suspension the mixture of the afore mentioned solvents was added. Then, the extraction was conducted by shaking liquid samples or filters containing suspension on the linear shaker within 1 hour. Afterwards, the extract was separated in glass classifier by flushing the walls few times or by decanting from the filter. Then, anhydrous sulphate sodium was added in order to bind to the remaining wastewater. Afterwards, the purification of extracts on silica gel in SPE Baker chamber was made. The purified extracts were concentrated under nitrogen stream to the volume of 2 ml. The prepared extracts were analyzed using gas chromatograph with mass spectrometry (Fisons) system. The time of analysis was equal to 50 minutes. The flow rate of samples was 1,5 ml/min. The recovery values were determined for individual PAHs, and then the average concentration of the following PAHs: naphtalene, acenaphtylene, acenaphtene, fluorene, phenanthrene and anthracenewas calculated. The mentioned compounds include from 12 to 14 carbon atoms in the particle or 12 hydrogen atoms. They have rather low octanol/water partition coefficient.

3 Results and discussion

The content of organic compounds in wastewater taken from the coke wastewater treatment plant expressed by COD was on the level of 1412 mg/L, whereas pH was equal to 6.8. After the process of oxidation COD decreased to the value of 1046 mg/L. The initial concentration of naphthalene ranged the value of 2270µg/L. The changes in the naphthalene concentration during studies are included in Fig. 1.

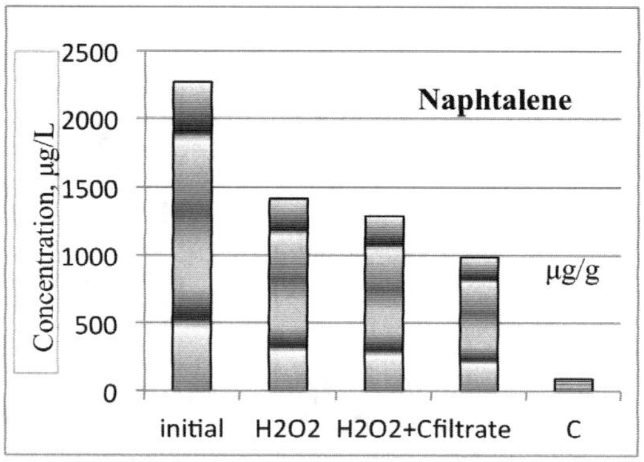

Fig. 1. Changes in the naphtalene concentration during oxidation and adsorption processes.

The efficiency of the naphtalene removal during oxidation and adsorption processes achieved 38% and 43%, respectively. The PAHs determination was carried out after filtering samples resulting in the isolation of dissolved fraction and suspension of activated carbon. The amount of naphtalene extracted from activated carbon was equal to 96µg/g C. The initial total concentration of 2-benzene ring hydrocarbons and one cyclopenthane ring ranged the value of 10523µg/L. After adding dihydrogen dioxide and PAHs oxidation the 77% decrease of hydrocarbons was noted. However, the efficiency degradation of individual compounds was diversified. The structure of acenaphtylene was stable resulting in the concentration before and after process only of 23% lower than the initial one. During simultaneous processes of oxidation and adsorption the decrease in the acenaphtylene content ranged 80%. Changes in the acenaphtylene concentration during oxidation and adsorption processes are given in Fig. 2.

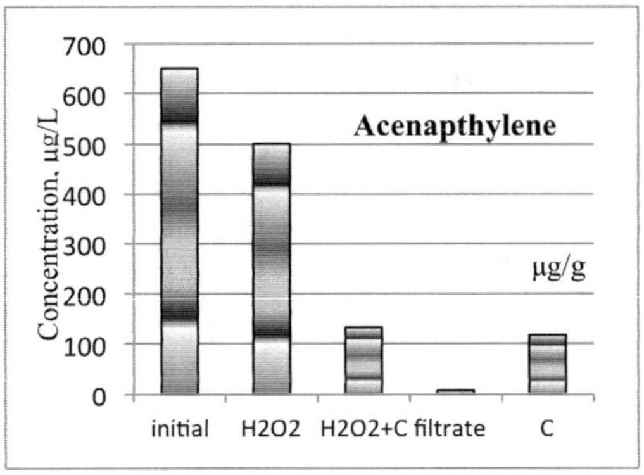

Fig. 2. Changes in the acenaphtylene concentration during oxidation and adsorption processes.

In Fig. 3 and 4 changes in the concentration of acenaphtylene and fluorine during oxidation process are included, respectively.

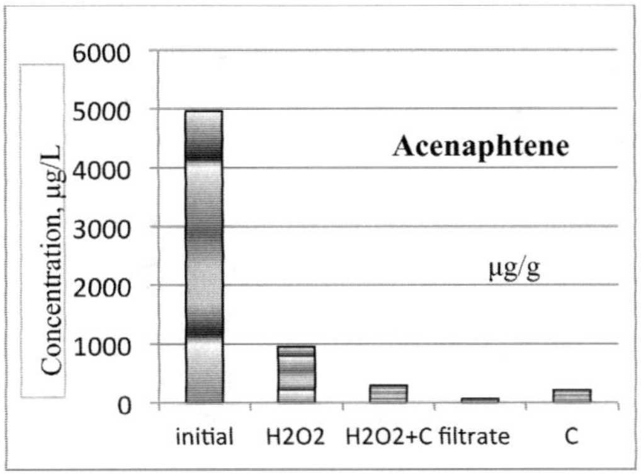

Fig. 3. Changes in the acenaphtene concentration during oxidation and adsorption processes.

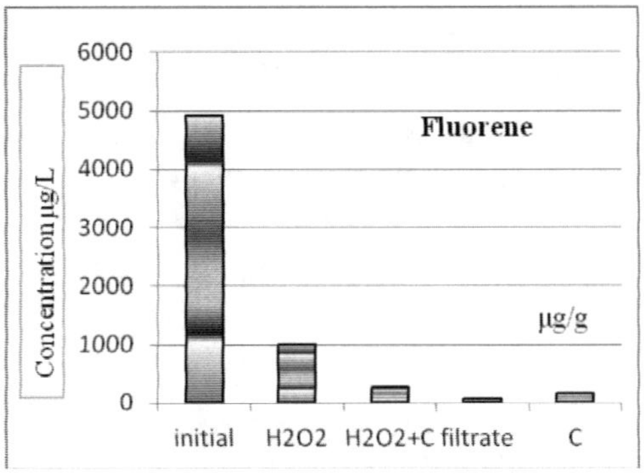

Fig. 4. Changes in the fluorene concentration during oxidation and adsorption processes.

In contrast to acenaphtylene, the content of acenaphtene and fluorene was lower than the initial one of 80-81%. In simultaneous adsorption and oxidation process the decrease of PAHs was higher than in the case of oxidation and ranged the value of 94%. The PAHs determination was carried out after filtering, isolating dissolved fraction and suspension of activated carbon. The total concentration of three compounds amounted 146µg/L and was lower of 98% from the initial one.

Extraction of PAHs from the suspension allowed to determine compounds that were adsorbed onto activated carbon. The concentration of three 3-ring compounds on activated carbon suspension ranged the value of 509 µg/g C. That indicates of the high importance of sorption despite the high solubility in the water. The total concentration of 3–ring hydrocarbons did not exceed the value of 3600 µg/L.

In Fig. 5 and Fig. 6 changes in the phenanthrene anthracene concentration during studies are given, respectively. After oxidation with the usage of dihydrogen dioxide the total concentration was lower than the initial one of 47%. The addition of activated carbon improved the efficiency of PAHs removal of 26%. In the filtered samples the PAHs concentration was lower than the initial one of 90%.

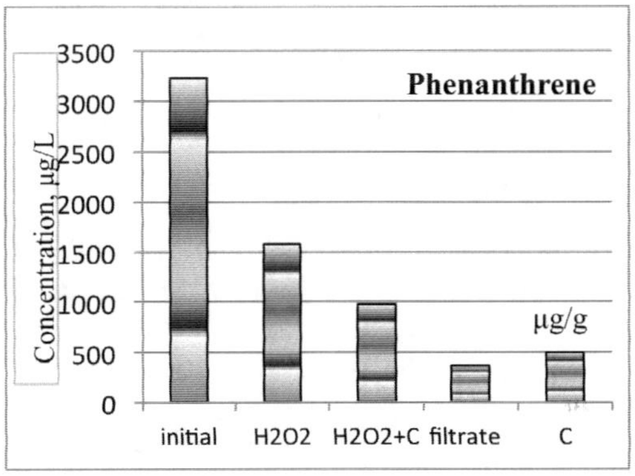

Fig. 5. Changes in the phenanthrene concentration during oxidation and adsorption processes.

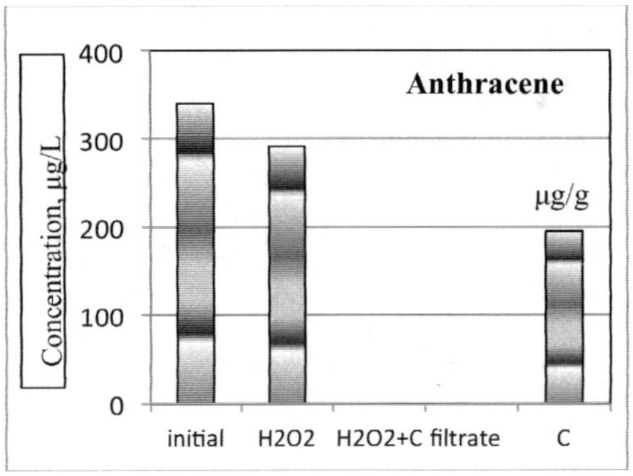

Fig. 6. Changes in the anthracene concentration during oxidation and adsorption processes.

From the activated carbon 700 µg/g C of PAHs were extracted. The results indicate that more favorable conditions of 2- and 3-ring hydrocarbons removal appeared during simultaneous oxidation and adsorption. It should be noted that under technological conditions of carrying studies the highest level of the decrease was found for hydrocarbons including cyclopenthane ring in the particle structure.

4 Conclusions

Based on the conducted studies the following conclusions can be formulated:

- The efficiency of degradation of studied hydrocarbons was in the range of 14- 81% under oxidation whereas, under simultaneous oxidation and adsorption onto activated carbon was in the range of 43-100%, respectively.
- Simultaneous oxidation and adsorption may be effectively apply in the additional treatment of wastewater including PAHs.

The studies were supported byBS-PB-402-301/11

References

1. Rozporzadzenie Ministra Środowiska z dnia 18 listopada 2014 r., Dz. U. 2014 poz. 1800
2. M. Włodarczyk-Makuła, *Physical and Chemical Fates of organic micropollutants* (Scholar-Press, 2015)
3. M. Włodarczyk-Makuła, Ann. Set Environ. Prot. **13**, (2011)
4. M. Włodarczyk-Makuła, A. Turek, *Monografie Komitetu Inżynierii Środowiska Polskiej Akademii Nauk* (Polska Akademia Nauk, 2012)
5. L. Dąbek, E. Ozimina, A. Picheta-Oleś, Annu. Set Environ. Prot. **15**, (2013)
6. M. Smol, M. Włodarczyk-Makuła, D. Włóka, CEER **13**, 2 (2014)
7. E. Ayranci, O. Duman, Chem. Eng. J. 156 (2010)
8. R.C.Brandli, T. Hartnik, T. Henriksen, G. Cornelissen, Chemosphere **73**, 11 (2008)
9. K. Barbusiński, *Aktualne Zagadnienia W Uzdatnianiu I Dystrybucji Wody*(Wydawnictwo Politechniki Śląskiej, Gliwice, 2013)
10. M. Smol, D. Włóka, M. Włodarczyk-Makuła, Water Air Soil Pollut. **229**, 154 (2018)
11. B. Pan, X. Baoshan, Environ. Sci. Technol. **42,** 24 (2008)
12. H.H. Huang, M.C. Lu, J.N. Chen, C.T. Lee, Chemosphere **51** (2003)
13. M.A. Nkansah, A.A. Christy, T. Barth, G.W. Francis, J. Hazard. Mater. **217** (2012)
14. M. Yuan, S. Tong, S. Zhao, C.Q. Jia, J. Hazard. Mater. **181**, 1 (2010)
15. F. Beltrán, J. Rivas, P. Álvarez, M. Alonso, B. Acedo, Ind. Eng. Chem. Res. **38** (1999)
16. A. Kornmüller, U. Wiesmann, Water Res. **37** (2003)
17. M. Trapido, Y. Veressinina, R. Munter, Environ. Technol. **16** (1995)
18. E. Neyens, J. Baeyens, J. Hazard. Mater. **B98** (2003)
19. S.R. Pouran, A. Abdul, R. Abdul, M. Wan, W.D. Ashri, J. Clean Prod. **64** (2014)
20. A.N. Soon, B.H. Hameed, Desalin. **269** (2011)
21. L.C. Toledo, A.C. Silva, R. Augusti, R.M. Lago, Chemosphere **50** (2003)
22. A. Rubio-Clemente, R.A. Torres-Palma, G.A. Penuela, Sci. Total Environ. **478** (2014)
23. A. Turek, M. Włodarczyk-Makuła, W.M. Bajdur, Desalin. Water Treat. **57,** 3 (2016)
24. A. Turek, M. Włodarczyk-Makuła, A. Nowacka, Ann. Set Environ. Prot. **16** (2014)

Impact of human activities on the quality of groundwater at Dobra intake in Jaworzno

Izabela Zimoch[1,*], and *Sylwia* Kręcichwost[2]

[1]Silesian University of Technology, Institute of Water and Wastewater Engineering, 44-100 Gliwice, 18a Konarskiego St., Poland
[2] Jaworzno Waterworks Ltd, 43-600 Jaworzno, 34 Św. Wojciecha St, Poland

Abstract. The paper identifies the causes of contamination and analyses changes in the quality of underground water resources at the Dobra intake, which supplies water for a separated group of Jaworzno inhabitants. The analysis takes into consideration such aspects as land conditions and their influence on physicochemical changes of water taken in in relation to the effective legal requirements for potable water. Moreover, an integral part of analysis of potential threats was an assessment of the effect of closed municipal waste disposal site and mining activity on the quantitative and qualitative resources of underground water in the area supplied by the Dobra intake in Jaworzno.

1 Introduction

Jaworzno is a city shaped by the growth of industry and urbanisation for hundreds of years. Former mining expansion and current widespread liquidation of mines, happening in Jaworzno as well, have influenced the quality and volumes of underground water resources. Strong human impact arising from mining activity conducted in the area of the city led to considerable transformations in the underground water environment. Both surface and underground mining contributed to this. During liquidation of surface mines, and in turn stoppage in functioning of drainage systems, workings became flooded and surface post-exploitation water bodies formed, which are currently used as recreation sites. An example of such human-related changes is the resource area of the Dobra intake in Jaworzno in connection with the formation of the Gródek quarry, which is currently a recreation a scuba-diving centre.
Apart from the mining activity, what had a considerable influence on the quality of underground waters in Jaworzno in terms of human impact, it was also the storage of waste and manner of use of waste disposal sites. The most popular method for disposal of municipal waste in Poland is still their storage, both compliant and not complaint with the currently effective legal requirements. Not long ago, the role of waste disposal sites was played by ground excavations, post-mining excavations, wild open pits, often without any safety measures [1].

Intense development of mining activity and its gradual extinguishing in the 1990s had a profound effect on the quantitative and qualitative dynamics of changes in the quality of underground waters in the resource area of the Dobra intake in Jaworzno. The result of those activities is two identified sources of threat for water quality at the Dobra intake, i.e.: municipal waste disposal site and the Gródek quarry.

The Dobra underground water intake takes water for supplying the inhabitants of specific districts of the city of Jaworzno. This is an underground water intake from Triassic formations with two bored wells D-1 and D-2. The confirmed volume of intake resources is 4.494 m^3/day (187.50 m^3/h). The direct surroundings of the intake are wastelands, farmlands (in the area of the wells) and compact urban development (in the area of the other facilities of the water treatment plant (i.e. WTP). Within 2 km from the intake, there are the following: to the southeast, the Gródek excavation after dolomite use (currently flooded in part), to the east, urban municipal waste disposal site (located in post-dolomite excavations) and to the northwest, a cemetery. 6 km to the northwest from the area in question there are vast railway areas of the Jaworzno-Szczakowa station. Ca. 5 metre from the D1 well, there is the Łużnik stream, being a right affluent of Kozi Bród. Identification of sources of underground waters contamination and monitoring of changes occurring in their quality is a considerable aspect from the point of view of safe management of water supply systems (WSS) [2-7]. WSS management based on implemented procedures of risk assessment is compliant with the amended Drinking Water Directive – 2015/1787 of 06.10.2015, L 260/6 – and Polish standard PN-EN 15975-2 Security of water supply.

2 Technical characteristics of the Dobra intake

Wells D-1 and D-2 of the Dobra intake were built in ca. 1917 and were used (mostly well D1) until mid 1987 with constant efficiency of over 1.000 m^3/day (over 42 m^3/h), with the determined depression of ca. 3 m [8].

In the area of the intake, the insulation properties of ground were heavily disturbed during surface

* Corresponding author: izabela.zimoch@polsl.pl

exploitation of raw rock materials, mainly limestones and dolomites, by Zakłady Dolomitowe Szczakowa and abandoned open pits after exploitation of sand, which led to considerable liquidation of soil layers, formation of sinks and unorganised waste disposal sites. Moreover, the formations of rocky substrate in that area exhibit weak insulation properties due to their natural porosity, karst formation and many faults [2]. Apart from adverse natural conditions, human impact effects are also visible: they result in depressurisation of orogenic belt with mining drainage. Exploitation activity is also connected with dewatering of mines, which results in depression funnels, which cover extensive urban areas in the case of the city of Jaworzno. What is also important is the inflow of underground waters after the exploitation ends and change in hydrological conditions in the areas of the funnels [9]. Inflowing mine waters have a characteristic, high mineralisation level, are characterised by high salinity, sulphate content and general hardness. Frequently, the effect of mine waters on underground resources makes it impossible to collect water for potting purposes, e.g. due to too high chloride concentration, as in the case in another underground water intake (WTP J. Dąbrowski) in Jaworzno.

Despite close proximity, waters taken from wells D-1 and D-2 have different qualitative content. In general, the quality of waters taken from well D-1 is better compared to well D-2, apart from winter and spring time, when colour in the water taken in by well D1 increases to values not accepted by the consumer. The water in D-2 well is characterised by periodically high contents of boron and lead. Parameters influencing decrease in the quality of the waters taken from well D-2 are also Cr^{6+} ions and sulphates (SO_4^{2-}) and their presence must be linked with the neighbourhood of municipal waste disposal site located in a closed quarry remaining after dolomite exploitation. This facility has had a considerable adverse effect on the quality of underground waters in the area of the Dobra intake.

3 Selected parameters characterising the quality of underground water from the area of the disposal site and the intake

Undoubtedly, what has had an influence on contamination of the environment in the area of supply of the Dobra intake has been the storage of municipal waste in the excavation site of former dolomite mine since 1971. The quality of that waste and the manner of their storage, as well as the very use of the disposal site have influenced the quality of soils and waters in that region. What caused serious threat to the environment were mainly activities connected with illegal storage of dangerous waste from the local tannery in the past. At the turn of 2002/2003, the influence of drains from the disposal site on the waters of the Dobra intake for content of chromium probably from tannery waste was analysed [10]. Attention was paid to the tannery waste deposited in the area of the disposal site in the 1990s, which waste contained a considerable concentration of chromium, which led to contamination of the vadose zone under the municipal waste disposal site and penetration of chromium to underground waters in the area of the Dobra intake. Chromium concentrations (Fig. 1 and Fig. 2) in the examined water samples varied from amounts characteristic of the geochemical background (< 0.005 mg/L) to concentrations of ca. 0.04 mg/L. The contents of that element exceeding 0.01 mg/L constantly occurred in piezometer P-II and in well D-2 of the Dobra intake in years 2000-2002. Analyses of dynamics of water chromium content illustrated in Fig. 1 show that in piezometer P-II water, the average annual chromium content decreased by 31% over 30 months to the level of 0.012 mg/L (September and October 2002).

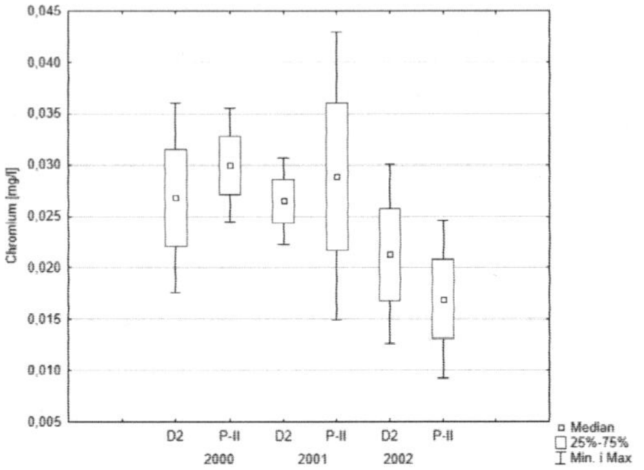

Fig. 1. Chromium contents in water samples taken from the piezometer P-II and the well D-2 in 2000-2002.

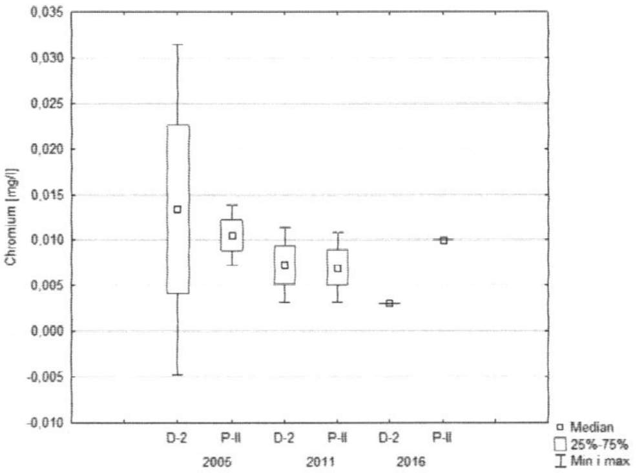

Fig. 2. Chromium contents in water samples taken from the piezometer P-II and the well D-2 in 2005, 2011 and 2016.

A decreasing trend in chromium content was also observed in well D-2 of the Dobra water intake, which from 0.035 mg/l measured in 2000 dropped to 0.015 mg/L. What was also observed was a correlation between seasonal changes in underground water table and chromium content in the examined water samples. The observed decrease in underground water table is related to lower chromium contents and – analogously – the rebuilding of underground water table is correlated with higher chromium contents. The observed phenomenon can be associated with more or less intense flushing of strongly contaminated vadose zone located directly under the disposal site, which results from high fluctuations in the ordinate of underground water table in that region [10]. The fact that the disposal site has an effect on the intake is confirmed by the results of monitoring of underground waters performed in the area of the disposal site later on.

Studies conducted in following years 2005-2016 showed that from 2010 there had been a considerable decrease in chromium content in underground waters of the Dobra intake both in piezometers and the D-2 well.

3.1 Microbiology

Over the 9 first years of use of the disposal site, liquid impurities, i.e. municipal sewage, were poured over the deposited waste periodically. Until 1978, the waters taken from the Dobra intake were compliant with the microbiological quality requirements for potable waters. In mid-1979, faces-type bacteria contamination occurred in the underground waters of the intake in question and in December of that year considerable bacteriological contamination took place [11]. Therefore, a sanitary inspectorate prohibited pouring sewage on the disposal site and ordered chlorination of the water from the Dobra WTP before supplying it to the consumers. Physicochemical analyses from the period of 14 years of use (1978-1991) showed that the highest contamination of the waters was observed in years 1986-87, when water level in the well was decreasing. A hydrogeological expert opinion prepared by the Expert Team of the Society of Mining Engineers and Technicians showed that the phenomenon in question was caused by dewatering of dolomite deposits by the Gródek mine, the wells of which drain the same water-bearing level of the Triassic. In the 1990s, during intensifying disappearance of water in the well of the Dobra intake, a decision was made to feed the well with waters from the dolomite mine dewatering barrier. Three wells and pumping systems directed the water from the local excavation to the well of the Dobra intake [11]. At that time, the microbiological quality of water was characterised by 400 CFU/mL of *Coli* group bacteria. The taken action decreased the problem of microbiological water contamination in the Dobra intake to 95%. Since 1992, laboratory analyses of underground water quality have shown single, incidental occurrences of *Coli* group bacteria and the overall number of microorganisms both in piezometers and wells D1 and D2.

3.2 Boron

Out of all microelements, boron (B) is the potential, characteristic contaminant for drains from municipal waste disposal sites. To remove it from water, it is necessary to apply some special methods like ion exchange or reverse osmosis. In the case of small, local potable water intakes, treatment with energy-draining and cost-draining methods becomes unprofitable, which – in the case of persistent excess of boron content in potable waters – puts an intake in danger of closing.

Persistent boron content in underground waters (Table 1) in studies covering the period from 2012 to 2016 confirm the hydraulic connection of the intake, mainly well D2, with waters from the drainage zone from the disposal site. Periodically, the water taken in well D2 does not meet the requirements of the Regulation of the Minister of Health (RMH) on quality of water intended for consumption by people. To meet the legal requirements for the quality of waters intended for consumption by people, the Dobra WTP uses mixing of streams of waters from well D1 and D2 in a proper ratio, which guarantees that boron content remains at a level lower than 1 mg/L.

Table 1. Boron contents in water samples taken from the piezometer P-I, P- II, P-III and the well D-1 and D- 2.

Date	Boron (mg/L) in 2012-2016					
	Well D1		Well D2		Piezometers	
	Norm	Result	Norm	Result	Norm	Result
2012	1	NDA*	1	NDA*	1	0.108
2013	1	0.27	1	1.39	1	0.19
2014	1	0.196	1	1.55	1	0.58
2015	1	0.19	1	1.50	1	0.11
2016	1	0.24	1	0.96	1	NDA*

*NDA- Not data avaliabe

3.3 Sulphates

The data illustrated in Table 2 show high sulphate content in underground water. The origin of sulphates in Triassic underground waters is difficult to pinpoint as they are polygenetic, can stem both from external and internal sources of contamination. The sources of external contamination are all types of solid waste disposal sites and liquid waste disposal sites, which are present in the area of the intake. On the other hand, the internal foci are products of natural geochemical processes occurring in the orogenic belt, but initiated by human activity [11-13].

Table 2. Sulphates contents in water samples taken from the piezometer P-I, P- II, P-III and the well D-1 and D- 2.

	Sulphates (mg/L) in 2004-2016		
Date	Well D1	Well D2	Piezometrs
2004	100.8	191.4	0.0
2005	99.7	180.3	106.7
2006	97.8	188.6	99.7
2007	89.4	199.1	100.0
2008	89.2	202.4	106.3
2009	101.0	225.6	0.0
2010	119.8	245.6	93.0
2011	84.0	184.0	78.9
2012	83.3	193.3	127.3
2013	97.8	167.4	87.5
2014	98.3	201.0	124.7
2015	93.1	177.2	101.4
2016	104.4	187.3	0.0

The Table 2 also shows differentiation in the quality of

waters taken from wells D1 and D2 despite their close proximity. Not only sulphates, but also parameters such as overall hardness, conductivity, boron, chromium and colour confirm considerable diversification in the content of those waters.

3.4 Colour

The above graph shows a diversification in waters taken from wells D1 and D2. The conducted analysis showed that in the 1st and 4th quarter there is the highest increase in colour of water taken from well D1 (maximum colour of 25 mg Pt/L on 17-03-2016), which is caused by thawing and long-term precipitation resulting in an abrupt increase in the water level in the local stream. Increase in colour during autumn, winter and spring time repeats year after year, which is illustrated in the above graph (Fig. 3 and 4). The conducted analysis showed that the most probable cause of abrupt increase in the colour, frequently above admissible values, shows that there is no insulation of the water-bearing layers in well D1. At the same time, infiltration was noted of underground waters from the Łużnik stream found nearby, which stream runs 5 metres away from well D1.

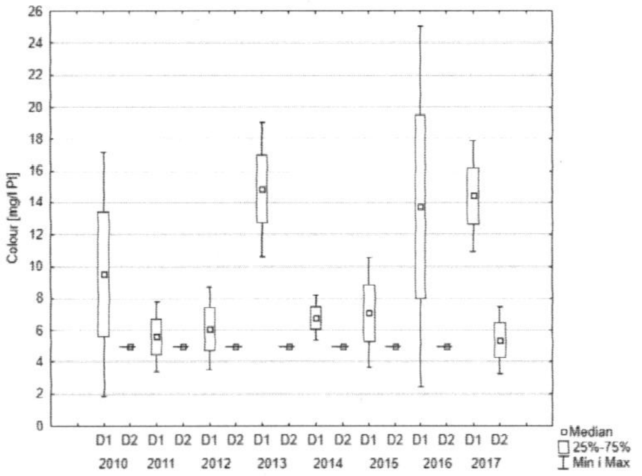

Fig. 3. Colour of water samples taken from the piezometer P-I, P- II, P-III and the well D-1 and D- 2 in 2010-2017 (I Quarter).

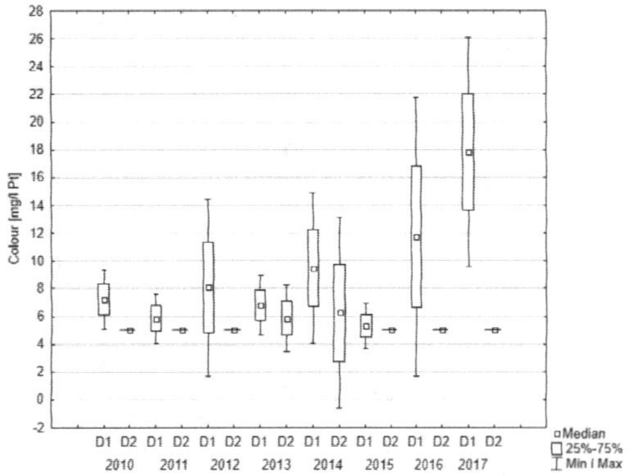

Fig. 4. Colour of water samples taken from the piezometer P-I, P- II, P-III and the well D-1 and D- 2 in 2010-2017 (IV Quarter).

4. Conclusions

Quality of underground waters in Triassic carbonate formations is monitored by three piezometers surrounding the closed waste disposal site and the Gródek quarry and two deep wells located in the area of the Dobra intake. The analysed parameters show a decrease trend in chromium content in water from the Dobra intake. Microbiological contamination of water has also stabilised compared to the 1990s.

What is currently a use-related problem of the Dobra intake is periodical failure to meet the requirements imposed by the Regulation of the Minister of Health of 07 December 2017 on the quality of water intended for consumption by people (Journal of Laws of 2017, item 1989), concerning intensified colour and boron.

Due to hydraulic connection of the underground water resources of the Dobra intake with the Gródek tank, there is high likelihood that the intake waters will be contaminated with impurities deposited at the bottom, formed in the process of flooding. For many years, at the bottom of the quarry there were two excavators, which could not be removed before cutting the power supply, and vehicles and an airplane flooded on purpose, to improve the appeal of the area as a scuba diving site. The former quarry is developing dynamically as a recreation and scuba-diving site, being the key threat of deterioration of the quality of underground waters intended for consumption by Jaworzno inhabitants.

At present, reclamation work is performed in the closed waste disposal site. In addition, constant, detailed monitoring of drains from the closed waste disposal site has been introduced. It aims to determine the level of threat for the quality of underground water resources of the Dobra intake and to develop and verify preventive actions for protection of those resources in accordance with the principles of water supply risk management.

Reference

1. J. Wiater, JEE **26**, (2011)
2. M. Działoszyńska-Wawrzkiewicz, K. Moraczewska-Majkut, K. E. Szymura, W. Nocoń, Ochr. Sr. **37**, 2 (2015)
3. M. Gono, M. Kyncl, R. Gono, I. Klosok-Bazan, PE **11**, 89 (2013)
4. J. Boguniewicz-Zablocka, I. Klosok-Bazan, V. Naddeo, Environ. Sci. Pollut. Res. (2017) (doi: 10.1007/s11356-017-0608-8)
5. M. Wolska, Water Sci. Technol. **71**, 4 (2015)
6. M. Bożym, I. Kłosok-Bazan, M. Wzorek, Pol. J. Environ. Stud. **27**, 3 (2018)
7. M. Wolska, Desalin. Water Treat. **52**, (2014)
8. B. Gajowiec, *Projekt badań hydrogeologicznych dla ujęcia „Dobra" w Jaworznie*, (Sosnowiec, 1998)
9. J. Pleczyński, JEE **7/8**, (1999)
10. Bauerek, M. Gwoździewicz, *Migracja chromu w wodach podziemnych węglanowej serii triasu rejonu ujęcia „Dobra" w Jaworznie*, (Główny Instytut Górnictwa, 2003)
11. M. Bacewicz, *Biuro Projektów Bud. Komun. w*

Krakowie: Dokumentacja – Ujęcie wód dolomitowych Dobra (1993)
12. J. Motyka, Z. Adamczyk, M. Czop, K. Dobyrn, Gosp. Sur. Min. **21**, 1 (2005)
13. M. Goździewicz, *Dodatek do dokumentacji hydrogeologicznej ustalającej zasoby eksploatacyjne wód ujęcia Dobra*, (2013)
14. C.B. Öman, C. Junestedt, Waste Manage. **10,** (2008)

Properties of catalyst for iron and manganese oxidation in filter materials

Joanna Jeż-Walkowiak[1,*], *Małgorzata* Komorowska-Kaufman[1], *Zbysław* Dymaczewski[1], *Marek M.* Sozański[1], and *Przemysław* Zakrzewski[2]

[1]Poznan University of Technology, Institute of Environmental Engineering, 61-131 Poznań, ul. Berdychowo 4, Poland
[2]Poznan University of Technology, Institute of Computing Science, 61-131 Poznań, ul. Piotrowo 2, Poland

Abstract. The aim of this work is to determine the properties of contact layers of grain of different type of filter media in term of their suitability for catalytical manganese oxidation and auto-activation for manganese removal from groundwater. The following oxidative filtration materials were investigated: auto-activated silica sand taken from full-scale filters (oxide coated filter media), Pyrolox – manganese ore – natural material and Hydrolit – artificially activated material. As a reference two natural not chemically active materials were tested: clean silica sand and clean chalcedonite. The investigation resulted in chemical composition and microstructure parameters of catalyst contact layers of grains. The modern analytical methods were used: EDAX, Raman spectroscopy, mercury and helium porozymetry. The data indicated that catalyst from autoacivated filtration material has better chemical composition and porosity parameters in comparison to tested catalytic filtration materials, making it more suitable for adsorption and catalytic manganese oxidation.

1 Introduction

Traditional technology of groundwater treatment consists of aeration followed by filtration. The iron and manganese removal occurs due to heterogenic and autocatalytic oxidation in catalytic-oxidative filtration beds. The catalyser is manganese dioxide – the product of Mn(II) oxidation.

Most common filtration material is silica sand. Clean fresh silica sand has no catalytic properties becoming an auto-activated material after covering grains with manganese oxides. This permanent oxides contact layer may be created with no chemical dosage [1,2]. The rippening of filtration material is often supported with biological oxidation process [3-6].

Another type of filtration materials are natural catalytic filter media (e.g. Pyrolox, Gabon manganese ore). This type of media is often use in small WTPs. Often these media loose in time the high efficiency in manganese removal and new steady state of dynamic balance has to be reached [7].

There are on the market artificially activated oxidative media – industrial products, manufactured with patented technology. It may be expected that the manganese oxidation catalysers of different filtration materials have different characteristics. The difference appear in chemical composition, microstructure parameters and polymorphic forms present compounds [8,9].

Manganese oxidation from the bivalent to quadrivalent form undergo on different pathways. One of the possible pathways of manganese oxidation is given in Fig. 1.

$[Mn(H_2O)_6]^{+2}$
↓
$Mn(OH)_2$
(Pyrochroite)
↓
Buserite
↓
δ-MnO_2
(Birnesite)
↓
γ-MnO_2
(Nsutite)
↓
β-MnO_2
(Pyrolusite)

Fig. 1. Manganese oxidation pathways [10].

Pyrolusite is the most stable form of manganese dioxide, It occurs naturally or is formed with use of chemical oxidants [6]. Pyrolusite has adsorption capacity to Mn(II) but does not have auto-catalytical properties [11]. Birnessite is the manganese oxide with the valency equal to 3.5 – 3.9, easily undergoes red-ox reactions, takes part in ion exchange, has high adsorption capacity for manganese, ensuring conditions for auto-catalytical oxidation [11]. Birnessite is a product of auto-catalytical

* Corresponding author: joanna.jez-walkowiak@put.poznan.pl

reactions of manganese oxidation taking place in systems of aeration and rapid filtration, without chemical dosage [2,11].

The manganese oxidation, taking place in filter media, is a heterogenic process on the surface of grains. An important factor affecting the adsorption of manganese on contact layer is the surface charge. The values of isoelectric point for quartz and manganese dioxide are shown in Table 1.

Natural groundwater usually has pH within the range of 5.5–7.5. At this pH surfaces of silica sand grains have negative charge. Higher negative charge in lower pH of birnessite (δ-MnO_2) than other manganese oxides results in higher potential ability for cations attraction.

Table 1. Point of zero charge for manganese dioxides [10,12].

Polymorphic variant of MnO_2	Formula	pH of isoelectric point
Hollandite	α-MnO_2	5.0
Pyrolusite	β-MnO_2	7.2
Nsutite	γ-MnO_2	5.6
Birnessite	δ-MnO_2	2.8
Quartz	SiO_2	2.0

The time and effects of auto-activation of media for manganese removal depend on the microstructure prameters of grains filter media. Investigation of specific surface and pore volume allows determination of the potential ability of the catalyser for manganese adsorption and further auto-catalytic Mn oxidation.

2 Goal and scope of the research

The aim of this work is to determine the properties of contact layers of grains in different type of filter media in term of their suitability for catalytical manganese oxidation and auto-activation for manganese removal from groundwater.

Following oxidative filtration materials were investigated (Table 2):
- auto-activated silica sand taken from full-scale filters (oxide coated filter media),
- Pyrolox – manganese ore – natural material,
- Hydrolit – artificially activated material.

As a reference two natural, not chemically active, materials were tested:
- clean silica sand,
- clean chalcedonite.

3 Methods

For catalytic contact layer of filter media grains investigation the following methods were used:
- energy dispersive spectroscopy – EDAX, coupled with SEM,
- Raman spectroscopy,
- mercury and helium porozymetry method.

Raman spectrometry was done with Renishaw apparatus. The SEM-EDAX analysis was done with Quanta FEG 250 (FEI) Microscope coupled with EDS Octane II Silicon Drift Detector (EDAX).

To determine the chemical composition of catalysts the energy dispersive spectroscopy EDAX, coupled with SEM was used. Weight shares of predominant elements were analysed using Microscope Quanta FEG 250 (FEI) with EDS Octane II Silicon Drift Detector (EDAX).

For the research on the identification of different polymorphic forms of oxides of the investigated catalysts, the method of μ-Raman spectroscopy was used. The non-polarised Raman spectra of filtration materials were recorded in the backscattering geometry using a Via Renishaw micro-Raman system. All spectra were recorded in the range of 100 to 2000 cm^{-1}. As an excitation light, the green line of argon laser operating at 514.5 nm (three accumulations of each spectra) was used. A laser beam was focused on the sample through the long working distance of the 50×/0.5 objective. The spatial resolution of the Raman spectra was about 2 μm. The power of the laser beam was less than 5 mW. All measurements were performed at room temperature. The crystalline structure of the catalytic contact layer in the manganese removal zone as well as silica sand beds were examined using the method of X-ray powder diffractometry (XRD). Examinations consisted of a comparison of diffraction diagrams obtained in Brucker D8Advance diffractometer for the analysed samples and the reference diagrams available in specialised databases [13]. The internal structure parameters pore volume and specific surface area was determined with the nitrogen adsorption method with a Tristar II 3020 apparatus.

Mercury porosimetry analysis is the progressive

Table 2. Element composition of tested materials.

Investigated material	Weight shares of elements, in %									
	Fe	Mn	O	Si	Ca	Al	K	Mg	S	P
Chalcedonite	0.6	–	61.5	33.3	0.36	3.54	0.32	–	0.32	–
Silica sand	0.17	–	60.7	37.4	–	1.42	0.24	–	–	–
Catalyser – iron removal zone	30.1	17.5	45.1	1.57	3.75	1.15	–	–	0.08	0.56
Catalyser – manganese removal zone	17.0	18.5	56.6	1.06	3.27	3.32	–	–	–	0.28
Pyrolox	3.66	48.95	40.6	1.84	–	2.73	1.52	–	–	0.07
Hydrolit	1.4	36.57	36.88	3.93	15.22	3.33	0.71	1.46	0.47	0.02

intrusion of mercury into a porous structure under stringently controlled pressures. From the pressure versus intrusion data, the instrument generates volume and size distributions using the Washburn equation.. Clearly, the more accurate the pressure measurements, the more accurate the resulting pore size data.

Mercury porosity analysis of tested materials was done using AutoPore9220 apparatus by Micrometrics. Applied instrument allows to determine the pore distribution in the range of pore diameter from 0.003 to 500 μm, with applied maximum pressure of 60000 psi. The helium porosimeter due to penetrating properties of helium allows to determine the submicropore structute of tested materials. It enables to precise measure of specific density of samples.

The sample of filtration material (of given mass-M and given volume V_{ext}) is placed into calibrated chamber of helium porosimeter and then helium is introduced. From the amount of introduced helium the volume of structure of filtration material - $V_{structure}$ is calculated. The total open porosity coefficient m_{open} is calculated according to the formula (1):

$$m_{open} = (V_{ext} - V_{structure}) / V_{ext} \quad (1)$$

Next the same sample is introduced into mercury porozymeter and the bulk density ρ_b is determined taking into account the volume of mercury – V_{Hg} (non-moisten liquid) driven out by the porous sample.

$$\rho_b = M/V_{Hg} \quad (2)$$

The dynamic porosity is calculated taking into account the volume of mercury introduced into the sample under pressure up to 60 000 psi. The calculation of pores volume includes only the volume of pores available for mercury: 0,01-360μm. The value of dynamic porosity is smaller than the value of open porosity.

4 Results and discussion

The photographs of investigated materials' surfaces using Raman spectrometer and scanning electron microscope show that the auto-activated contact layer is not homogenous and consists of centres of iron and manganese oxides.

The comparison of not-chemically active materials shows the more developed surface structure of chalcedonite than silica sand. This creates better suitability of chalcedonite for autoactivation for manganese removal from groundwater [2].

In Table 2 the element composition of investigated media are presented. The data were obtained using apparatus EDAX [14].

The EDAX results (Table 2) indicate, that:

- silica sand and chalcedonite are chemically homogenous, mostly consisting of SiO_2,
- catalyser of iron removal zone consists of iron (30.1%), manganese (17.5%), oxygen (46.0%), calcium (3.75%), indicating presence of iron and manganese oxides in the mass of catalyser.
- catalyser of manganese removal zone consist of manganese (18.5%), iron (17%), calcium (3.25%). The presence of iron and calcium in this catalyser lowers catalytical activity of contact layers.
- in Pyrolox manganese dioxide is the main constituent,
- Hydrolit is a mixture of manganese iron, calcium and magnesium compounds.

For identification of compound structure of investigated media the method of μ-Raman spectroscopy was used. The method is based on comparison of tested materials spectra and reference spectra. In Table 3 the values of characteristic wavenumber for iron, manganese and quartz oxides expected in contact layers are presented. For the investigation the Renishaw spectroscope was used. Raman spectra of tested materials and catalysers are presented in Fig. 2 – 7.

The Raman spectrum of the chalcedonite and silica sand samples, presented in Figures 2 and 3 respectively have the following Raman bands: 127, 206, 464 cm^{-1}. The same values of characteristic wavenumbers can be found in the reference spectrum of quartz (Table 3). Silica sand and chalcedonite are filtration materials with the structure of pure quartz.

Comparison of Raman spectrum for the iron removal zone with the reference plot of characteristic wavenumber (Table 3, Fig. 4) of iron oxides, indicated the presence of hematite αFe_2O_3 and maghenite γFe_2O_3 [15].

In manganese removal zone the presence of birnessite, was proved (Table 3, Fig. 5).

Table 3. Raman shifts for silica and iron and manganese oxides [11,15,16].

Compound	Wavenumber, [cm^{-1}]					
Quartz, SiO_2	127	206	464			
Maghemite γFe_2O_3	350	512	665	730		
Hematyte αFe_2O_3	265	300	345 – 395	515	645 – 670	715
Birnessite δ-MnO_2	495 – 505	570 – 575	635 – 655			
Pyrolusite β-MnO_2	538	665				
Nsutite γ-MnO_2	379	491	520	572	631	738
Ramsdelite	294	518	580	630	740	

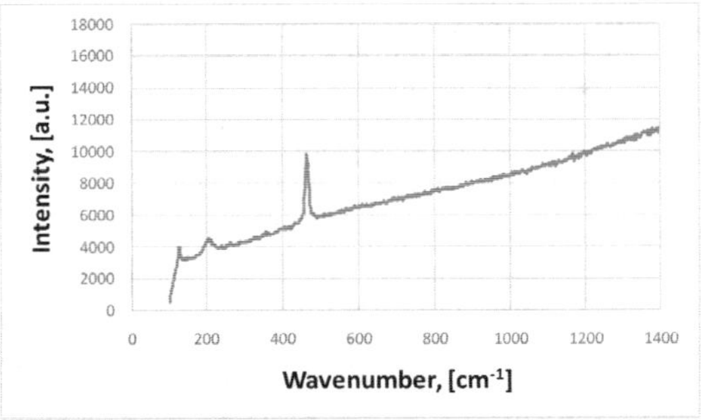

Fig. 2. Raman spectrum for chalcedonite.

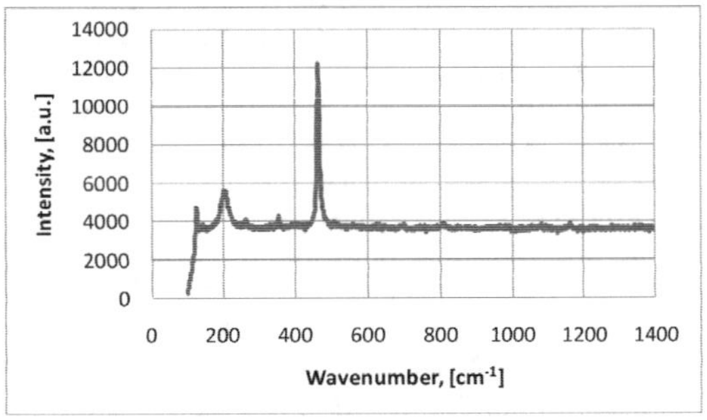

Fig. 3. Raman spectrum for silica sand.

In Pyrolox the presence of manganese oxides in the form of pyrolusite, ramsdelite and nsutite (Table 3, Fig. 6) were found.

Raman spectrum of Hydrolit indicated manganese dioxide in the form of pyrolusite (Table 3, Fig. 7).

In Table 4 porozymetry parameters obtained with mercury porozymetry method and helium porosity method are presented. AutoPore IV 9500 used in investigation allows to determine the pore structure within the range of pore diameter 0.003 – 500 μm.

Manganese removal zone catalyser has the high value of specific surface area 124.75m^2/g. The low value of bulk density 0.652 g/cm^3 correlates with high value of total porosity 80.552%. Established porosity parameters allow to consider manganese removal zone catalyser as a porous material with good adsorption properties. The specific surface area of Pyrolox and Hydrolit are few times lower than manganese removal zone catalyser.

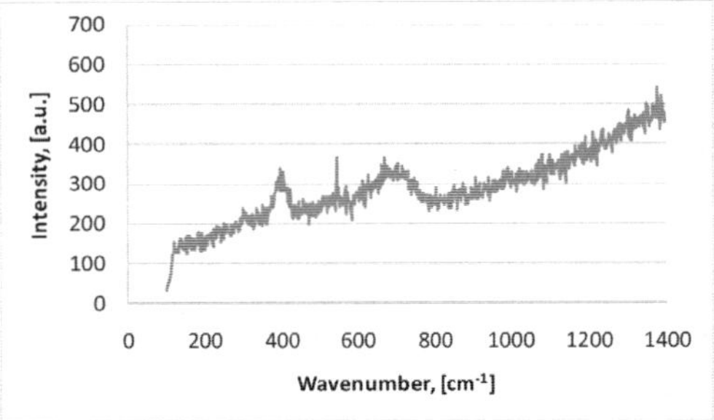

Maghemite

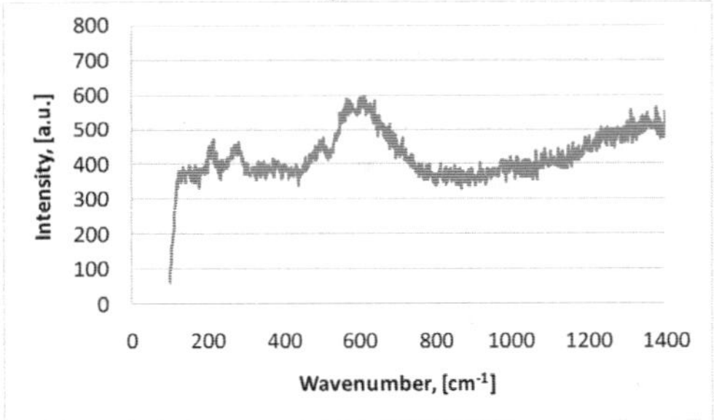

Hematyte

Fig. 4. Raman spectra for iron removal zone.

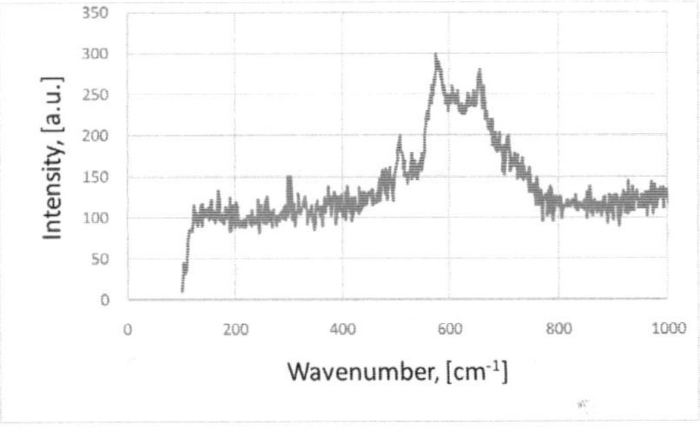

Fig. 5. Raman spectrum for manganese removal zone.

The lowest value of specific surface area have silica sand. On the other hand the specific surface area of chalcedonite is almost two times higher than Pyrolox and Hydrolit and 200 times higher than silica sand. It gives the evidence of better properties of chalcedonite to auto-activation for manganese removal from groundwater [2].

Table 4. Porozymetry parameters obtained with AutoPore apparatus.

Material	Mass density	Total open porosity	Bulk density	Dynamic porosity	Specific surface area
	[g/cm^3]	[%]	[g/cm^3]	[%]	[m^2/g]
Manganese removal zone of rippened silica sand bed	3.335	80.552	0.652	72.056	124.75
Pyrolox	4.03	23.81	3.08	20.35	2.58
Hydrolit	2.92	13.01	2.49	11.94	2.06
Silica sand	2.64	5.15	2.48	5.00	0.02
Chalcedonite	2.63	35.68	1.71	29.73	4.74

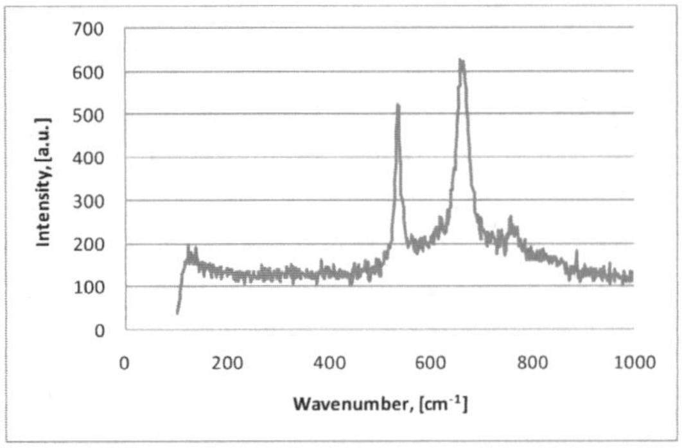

Pyrolusite

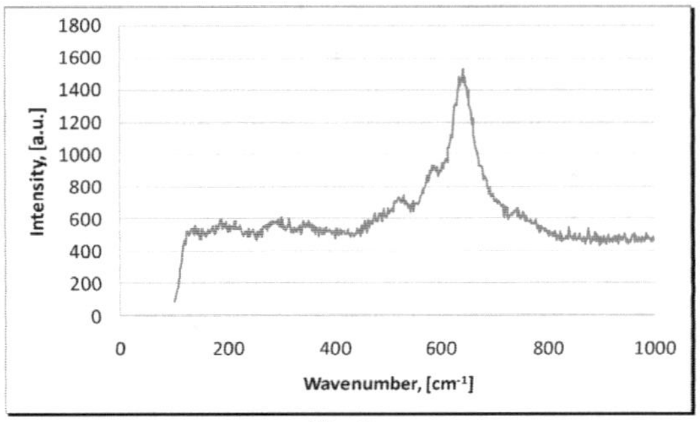

Nsutite

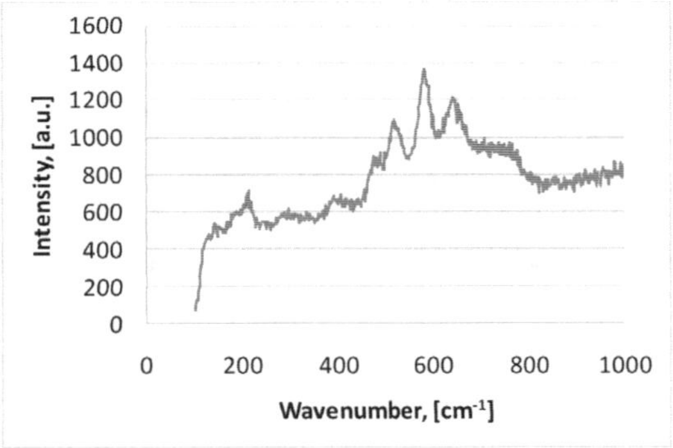

Ramsdelite

Fig. 6. Raman spectra for Pyrolox.

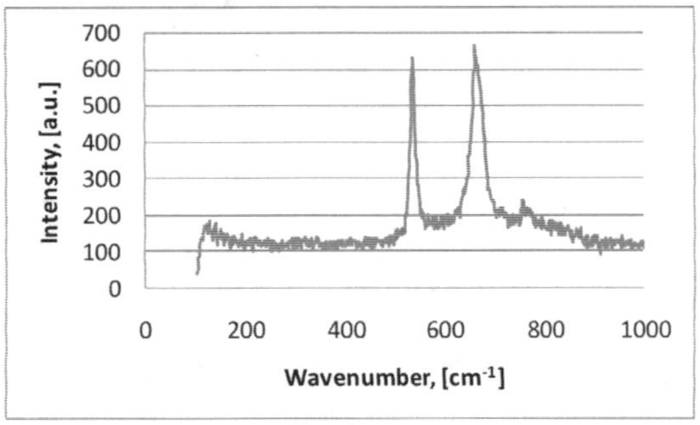

Pyrolusite

Fig. 7. Raman spectrum for hydrolit.

5 Conclusions

In the described investigation on catalysers for catalytic manganese oxidation, the methods of instrumental analysis were used. The obtained data allowed:

- determination of elements' shares and chemical composition of the investigated catalysers,

- determination of microstructure parameters e.g. specific surface area and porosity of the contact catalytical layer of tested media.

The analysis of catalyser of manganese removal zone pointed that this material is the most capable for effective manganese removal from groundwater in the process of heterogenic autocatalytic oxidation in filter bed. This capability comes from chemical composition of catalyser from auto-activated beds and because of high content of manganese dioxide in the form of birnessite, $MnO_{1.3\text{-}1.9}$ – not fully oxidized manganese dioxide with deficiency of oxygen. Such oxide becomes a semi-conductor able to transport electrons, guarantying good efficiency of heterogeneous reaction.

Catalyser from iron removal zone contains the iron oxides in the form of αFe_2O_3 – hematite and γFe_2O_3 – maghemite.

The analysis of microstructure also proof the capability of auto-activated catalysers for effective adsorption of manganese from treated water due to high values of the specific surface area and pore volume, measured by Autopore apparatus. Catalyser of the manganese removal zone has a specific surface area 124.75 m^2/g and total porosity 80.55%.

In Pyrolox the manganese dioxides were detected in the form of ramsdelite, nsutite and pyrolusite. This bed is characterised by much lower specific surface area 2.58 m^2/g in comparison to auto-activated catalyser.

Artificially activated Hydrolit is a mixture of manganese, iron, calcium and magnesium compounds. This bed is characterised by the low specific surface area of 2.06 m^2/g determined with mercury porosimetry method.

Investigation of two not-chemically active materials demonstrate the higher chalcedonite potential for manganese removal auto-activation due to higher, in comparison to silica sand, values of microstructure parameters.

Further research on relationships between the characteristic of catalysers and technological effects of groundwater treatment should be conducted. It will help the evaluation of mechanisms of heterogenous, catalytic manganese oxidation in filtration bed.

The authors would like to express their thanks for the financial support from research project 01/13/DSPB/0857.

References

1. J. Jeż-Walkowiak, Z. Dymaczewski, J. Wat. Sup. Res. Tech. AQUA **61**, 6 (2012)
2. J. Jeż-Walkowiak, *Wpływ właściwości złóż filtrów pospiesznych na efekty technologii odżelaziania i*

odmanganiania wód podziemnych (Wyd. PP, Poznań, 2016)
3. P. Mouchet, AWWA J. **84**, 4 (1992)
4. K. Olańczuk-Neyman, R. Bray, Pol. J. Env. Stud. **9**, 2 (2000)
5. R. Bray, K. Olańczuk-Neyman, Wat. Sci. Tech. Wat. Sup. **1**, 2 (2001)
6. R. Bray, *II Congress of Environmental Engineering, Lublin, Poland, Monografie PAN, vol.32*, (2005)
7. J. Jeż-Walkowiak, Z. Dymaczewski, A. Szuster-Janiaczyk, A.B. Nowicka, M. Szybowicz, Water **9**, 498 (2017)
8. J. Jeż-Walkowiak, *Intensyfikacja procesu odżelazia-nia odmanganiania wód podziemnych w złożach oksydacyjnych filtrów pospiesznych* (PhD Th., 2000)
9. J. Jeż-Walkowiak, Z. Dymaczewski, Ł. Weber, J. Wat. Sup. Res. Tech. AQUA **64**, 1 (2015)
10. W. Stumm, J.J. Morgan, *Aquatic Chemistry, Chemical Equilibria and Rates in Natural Waters, - Third edition* (Wiley Interscience, 1996)
11. J.H. Bruins, B. Petrusevski, Y.M. Slokar, J.C. Kruithof, M.D. Kennedy, Des. Wat. Tr. **55**, 7 (2015)
12. M. Kosmulski, *Surface charging and points of zero charge*, (CRC Press, Tylor&Fracis Group, 2009)
13. S. Jędrychowska, *Możliwości wykorzystania spektroskopii ramanowskiej w branży naftowej, cz. I*, (Nafta-Gaz, 2012).
14. A. Barbacki, *Mikroskopia elektronowa*, (Wyd. PP, Poznań, 2007)
15. M. Szybowicz, M. Koralewski, J. Karoń, L. Melnikova, Acta Phys. Pol. O. Acc. **127**, 2 (2015)
16. C. Julien, M. Massot, R. Baddour-Hadjeans, S. Franger, S. Bach, J.P. Pereira-Ramos, Sol. State Ion. **159** (2003).

Experiment in infiltration studies as a water treatment process

Sylwia Kołaska[1], *Joanna* Jeż-Walkowiak[1,*], and *Zbysław* Dymaczewski[1]

[1]Poznan University of Technology, Institute of Environmental Engineering, 61-131 Poznań, ul. Berdychowo 4, Poland

Abstract. This article contains the concept of research methodology of the water infiltration process. The described research was conducted at Dębina intake in Poznan, Poland. Based on the temperature variation curves, it was possible to determine the water retention time in the ground during the flow of water from the infiltration pond through piezometers to the collecting well

1 Introduction

The infiltration process for surface water treatment is often taken into account because of the high efficiency of this process and economic considerations. In the infiltration process without use of chemical reagents, we obtain a high degree of water treatment [1-6]. Characteristic pollutants of surface waters - such as suspension, organic compounds, pharmaceuticals bacteria, *Cryptosporidium* or cyanobacteria toxins are being removed [7-9] and the quality of water after the infiltration process is comparable to the quality of groundwater [4-6]. Comparing the complexity of surface water treatment systems with groundwater treatment systems and considering deficiency of groundwater at the same time, it is appropriate to practice artificial infiltration process [10,11]. Infiltration process might be consider as a sustainable process, effectively reducing the amount od reagents applied at water treatment plant.

The paper presents research conducted on the area of Debina infiltration intake in Poznan, Poland. Intake was chosen due to location and long term proper operation.

2 Brief description of Debina intake

Debina infiltration intake is located by the warta river between 247 to 251 km. There are located three rows of wells. The first row of wells is 80 m from the river, the second row of wells is 260 m from the river and the last is 340 m. Wells located directly by the river collect water from riverbank infiltration and artificial infiltration, next two rows of wells collect water only from artificial infiltration. Wells collect water from a depth of 10 – 18 m. The best water quality is obtained from the wells located 260 m from the river. In this localization water comes only from artificial infiltration and is separated from pollutions coming from outside the intake. The current capacity of debina intake is about 80 000 m^3/d. At present, modernization works are being performed. The modernization aims to restore the original intake capacity, achieved before a2 motorway construction – 100 000 m^3/d. The balance of supply of debina intake is [1,2]:

- artificial infiltration through infiltration ponds 65 – 76%;
- riverbank infiltration 16 – 27%;
- groundwater inflow 8 – 12%;
- rain water and melting waters – a few percent.

3 Methods and scope of research

Presented data are from preliminary work which had been done at debina intake. The research was conducted weekly. Measurement points were located along the second row of wells. To determine the water retention time in the ground, the following points were used: pond 3a, piezometers: i r./6, i r./7, well 37/ii, piezometers: i r./8, i r./ 9, pond 9 (Fig. 1). Piezometer i r./6 is located about 15 meters from the pond 3a. Then there is piezometer i r./7 at a distance of 40 meters from piezometer i r./6. The distance to well 37/ii from piezometer i r./7 is about 24 meters. From the other side we have piezometer i r./9 at a distance of about 14 meters from the pond 9. Another piezometer i r./8 is about 44 meters away from piezometer i r./9. Well 37/ii is about 24 m from piezometer i r./8 (Fig. 1).

An important parameter measured directly at debina intake was water temperature. Based on the determined temperature variation curves, it is possible to determine the water retention time in the ground during the flow of water from the pond through piezometers to the collecting well. In wells the temperature was always measured at the same depth. When it comes to piezometers, the temperature was measured every meter, starting with the first full meter below the dynamic water level, and ending with the last full meter before the bottom of the piezometer.

In research work, this non-invasive method of determining retention time was selected. It was mainly chosen to eliminate any interference in the work of the intake and to simplify carrying out measurements.

To determine the water quality change during the

* Corresponding author: joanna.jez-walkowiak@put.poznan.pl

passage through the ground the chemical analysis in water sample taken from ponds, piezometers and wells was done. Three series of samples were taken for laboratory analysis. In the facilities from which water was collected for analysis, pumping was always carried out at least twice to replace the volume of water before taking the sample.

4 Results and interpretation

4.1 Determination of water retention time in the ground

Water retention time in ground from pond 3a to piezometer I r./6 was assessed for 3 days, according to the water temperature drop after the decrease of air temperature. The water temperature drop took place 3 days after the air temperature decrease.

The retention time of water at the distance from the piezometer i r./6 to i r./7 was estimated at about 3 weeks. In the piezometer nearer to the pond, the water temperature rises from 25.04.2013 and in piezometer i r./7 (Fig. 2) 17.05.2013 – 3 weeks later (Fig. 3). In addition, there can be seen characteristic pattern of water temperature drops in piezometer i r./6 14.02.2013, which results in a drop in water temperature 07.03.2013 in piezometer i r./7, which also gives a three week retention time.

Comparing the air temperature in November 2012 (Fig. 2) with the course of the water temperature curve in piezometer i r./7, it was noted that the temperature drop of 14.12.2012 in the piezometer was most likely due to lower air temperatures in the period of 15–18 November. After these days, the temperature of the air increased as it was in case of temperature distribution of the piezometer. It is less than 4 weeks. Knowing that a few days is retention to the first point of measurement, estimated three weeks of retention between them are correct.

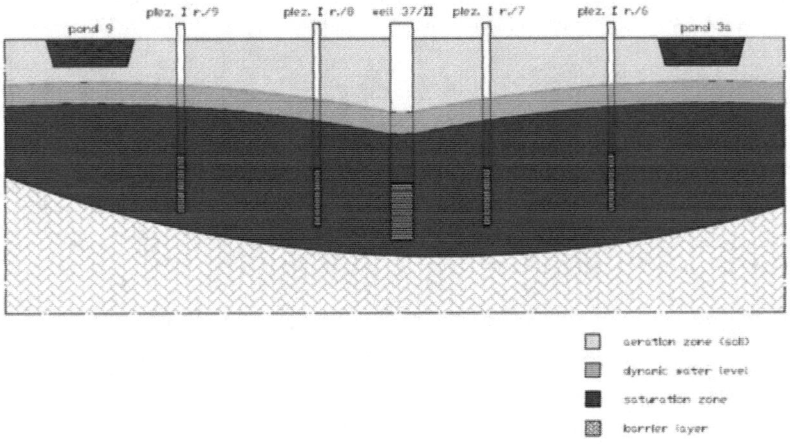

Fig. 1. Scheme of arrangement of measurement points.

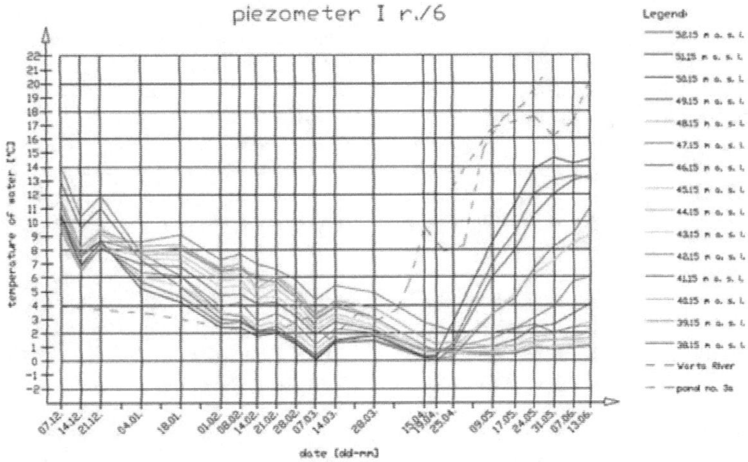

Fig. 2. Temperature of water in the piezometer I r./6.

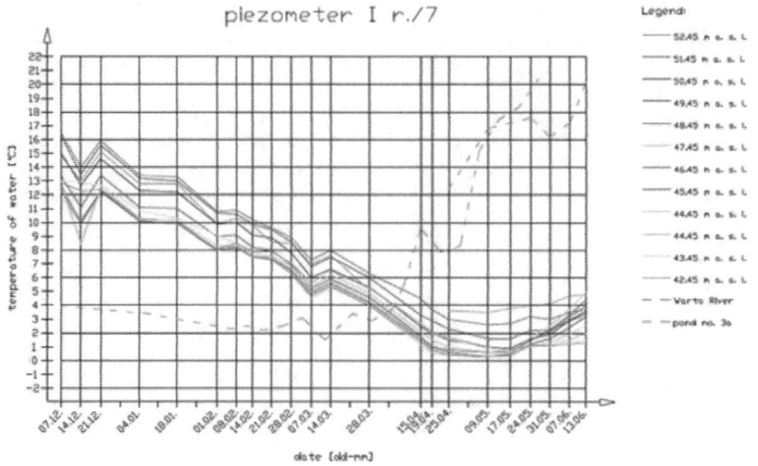

Fig. 3. Temperature of water in the piezometer I r./7

Water retention time between pond 9 and piezometer I r./9 is, as in the case of remaining piezometers located directly at ponds, several days. Water retention between two piezometers I r./9 and I r./8 takes about two weeks. Water flow time from piezometers to well 37/II was estimated to be 7 weeks.

4.2 Variability of chemical parameters of water

Under the influence of the infiltration process, surface water introduced into the ground through the infiltration ponds acquires properties similar to groundwater. The most representative results of the changes in water quality were found to be variations in the concentration of iron in the infiltration water.

The analysis of iron concentration in water samples was carried out with respect to the sequence of measurement points, from the ponds to the wells, starting with the shortest ground retention time. Presented results come from two measurement series made on 19.04.2013 and 24.05.2013.

To illustrate the change of iron concentration in water during the ground passage the flow line presented in fig.1 is chosen.

The largest increase in iron content is noted in the first phase of water flow from ponds to piezometers (from pond 3a to piezometer I r./6), where the average increase is over six times. Then the increase in iron concentration loses its dynamics.

The water flow time from the pond 9 to the piezometer I r./8 was 2 to 3 weeks. Water from pond 9 to the piezometer I r./8 flows mainly in the saturation zone and is filtration (remember that water called infiltration water actually is infiltration water and filtration water). As can be seen, the increase in iron concentration on this flow path is slightly smaller compared to the increase between pond 3a and piezometer I r./6.
The results are presented in Fig. 4.
Being analyzed in three measurement sequences, the infiltration (infiltration and filtration) process can be summarized as a factor that significantly increases the concentration of iron in water.

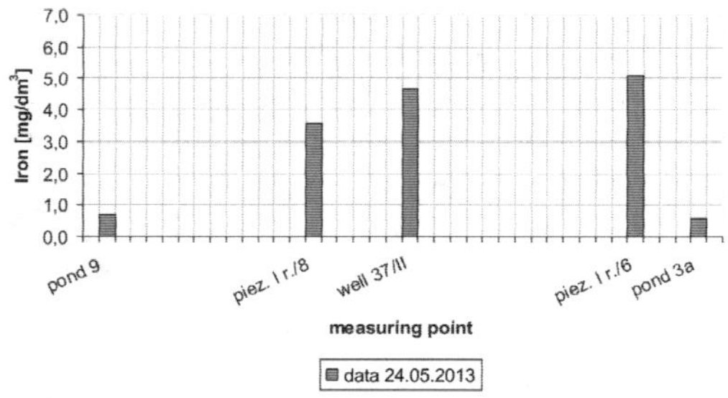

Fig. 4. Change of iron concentration in water by infiltration.

5 Conclusions

The research was carried out as a preliminary study, allowing to recognize the process conditions prevailing at Debina intake in Poznan. The methodology used to measure water temperature changes proved to be useful for determining the retention time of infiltration water in the soil. The method is a not chemical way of doing research which does not interfere with the work of the infiltration intake.

The conducted research on the area of the infiltration intake allowed to determine the water retention time in the selected infiltration water flow sequences at Debina intake.

Analysis of iron concentration confirmed the tendency of this parameter to change in infiltration water. The results confirm the existence of a relationship between the retention time and the efficiency of the process.

Recognition of process conditions on the area of Debina infiltration intake allowed exact planning of a further research works with an extended network of measuring points.

The authors would like to express their thanks for the financial support from research project 01/13/DSPB/0857.

References

1. Grunheid at al., Wat. Res. **39**, (2005)
2. A.K. Thakur C.S.P. Ojha, Int. J. Sediment Res. **25**, (2010)
3. Hoffmann A., G. Gunkel, Limnologica **41**, (2011)
4. Bartosik, I. Chomicki, T. Jankowski, Ochr. Środ. **3**, (2007)
5. Bartosik, I. Chomicki, T. Jankowski, *Quality changes of Warta River waters after infiltration process at Debina intake in Poznan* (Materials from company AQUANET SA, Poznan, Poland, 2007)
6. *Waterworks and sewerage in Poland – tradition and present days* (Edit. Polska Fundacja Ochrony Zasobów Wodnych, Poznań – Bydgoszcz 2002)
7. M.J. Miller at al., Wat. Res. **35** 6 (2001)
8. D.W. Metge at al., Wat. Res **44** (2010)
9. T. Heberer at al., Ground Water Monit. Rem. **24**, 2 (2004)
10. C.J. Lorenzen at al., Environ. Earth Sci. **61** (2010)
11. R.D. Rakesh at al., Hydrogeol. J. **18** (2010)

Application of modified concrete to remove surfactants from water

Kamil Kuźmiński[1], *Antoni W.* Morawski[2], and *Magdalena* Janus[1,*]

[1]West Pomeranian University of Technology, Faculty of Civil Engineering and Architecture, Department of Sanitary Engineering, al. Piastów 45, 70-311 Szczecin, Poland
[2]West Pomeranian University of Technology, Faculty of Chemical Technology and Engineering, Institute of Inorganic Chemical Technology and Environment Engineering, ul. Pułaskiego 10, 70-322 Szczecin, Poland

Abstract. In these studies two types of modified concretes in photocatalytic mineralization of anionic surfactant sodium dodecylbenzenesulfonate (NaDBS) were used. Commercial TiO_2-P25 and nitrogen modified TiO_2-N as photocatalysts for concrete modification were employed. The studies were conducted in the special designed reactor which was a simple model of the secondary settler used at the sewage treatment plant. The mineralization of anionic surfactant NaDBS after 18 h of UV light irradiation amounted to 48% for concrete modified by TiO_2-P25 and 37% for concrete modified by TiO_2-N.

1 Introduction

Active surface agents are the group of organic compounds most commonly used in various industries, services and households. In the industry, they are used for washing and dyeing (textile) as wetting substances - for example: for foundry molds (metallurgy), in the production of pesticides, synthetic resins, paints and varnishes (chemical industry), plastics, cellulose and paper, food (food industry) and moreover in printing , photography, cosmetics, pharmacy, zootechnics and medicine. In households surfactants are mainly used as washing agents [1-3]. The market research institute Ceresana Research expects the global active surface agents market to generate revenues of more than 41 billion US dollars in 2018 – translating to an average annual growth of 4.5% [4].

Surfactants can be diveded on: ionic (cationic, anionic), non-ionic and amphoteric, depending on their structure. Structure of surfactants consisting of hydrophilic part (ionic or strongly polar) and hydrophobic part, which is usually a hydrocarbon chain [5]. One of the most commonly used anionic surfactants are linear alkylbenzenenesulphonates (LASs), introduced in the 1960s to substitute the hardly biodegradable tetrapropylbenzenesulfonate. In plants, which are treating households wastewater mainly, LAS concentration under 10 mg/L can be found [6]. This concentration can expressly increase when municipal wastewater are also receiving wastes from industrial washing processes [7]. Some works has been reported to estimate the biodegradability and environmental safety of LASs directly and their intermediates degradation products [8]. As biodegradability has been proven, linear alkylbenzenenesulphonates removal from wastewater has generally been limited to conventional processes, like biological treatments. Percent of LAS removed in popular biological systems is usually between 90 and 95% by activated sludge and almost 70% by biological filter wastewater treatment [9]. However, biological treatment make some difficulties. For concentrations over 20 mg/L the pH self-regulation of the wastewater decreases and, as a consequence additional neutralization is required [10]. Moreover, in biological processes long retention time is needed to accomplish surfactant removal [9] which may create the need to reduce an amount (or time) of others conventional treatment methods (physical and chemical treatment, chemical oxidation).

An alternative to the conventional methods are the advanced oxidation processes (AOPs). Final products of surfactants decomposition by AOPs are CO_2, H_2O and inorganic residue. One of best known AOP method is heterogeneous photocatalysis based on semiconductors, such as most popular titanium(IV) oxide (TiO_2).

The role of the electron acceptor in the photocatalysis process perform O_2 forming in reduction reaction reactive oxygen radical $O_2^{\bullet-}$. Pair electron-hole which is photogenerated on the catalyst surface, may serve formation of the hydroxyl radicals by the oxidation reaction of the adsorbed molecules H_2O or OH^- [11] or can lead to oxidation of the adsorbed organic compounds [12]. An Amount of •OH radicals produced on the catalyst surface (surface hydroxylation) [13] is one of the important methods for determining its photocatalytic potential [14] and is next to adsorption (electron hole oxidation), significant oxidation way in whole photocatalysis process.

One of the methods of using the titanium dioxide is to use it to create a composite of common building

* Corresponding author: mjanus@zut.edu.pl

materials, such as gypsum [15] or concrete [16] with self-cleaning properties by its superhyrophilic surface. This is a simple method that does not require additional work and extra costs in the future. These photoactive materials can be effectively use against air or water pollution while retaining their original mechanical properties or even improving them [17].

In this study, the possibilities of sodium dodecylbenzenesulfonate (NaDBS) surfactant remove from water were investigated. For this purpose, concrete enriched with commercial and new, modified with nitrogen titanium dioxide under the UVA irradiation in laboratory scale were tested.

2 Materials and methods

2.1 Materials

An anionic surfactant from a group of LAS, sodium dodecylbenzenesulfonate (Sigma-Aldrich Company, Germany) as model surfactant was used. The molecular mass of this compound is 348.48 g/mol. The critical micellar concentration (CMC) in room temp. equal 554 mg/L [18].

Nitrogen modified TiO_2 (TiO_2/N) were obtained at the Institute of Inorganic Technology and Environment Engineering, West Pomeranian University of Technology in Szczecin, in the AutoLab installation E746 (HEL Ltd). 600 g of TiO_2 (Azoty, Police Chemical Factory S.A.) were placed reactor in 350 ml of 2.5% solution of NH_4OH. The modification was carried out at 100°C for 4 h. The photocatalyst was dried for 4 h at 100°C in an air atmosphere. Reference TiO_2 material was Aeroxide P25 (Evonik Company, Germany). Detailed information about semiconductors are presented in Table 1.

Table 1. Characteristic of used photocatalysts.

Photo-catalyst	S_{BET} [m^2/h]	avg molecular size [nm]	phase content	pH_{PZC}
TiO_2/N	345	190	67% amorphous, 30% anatase, 3% rutile	6.7
TiO_2-P25	56	26	75% anatase, 25% rutile	5.9

The concrete (WB-80, Cekol) was used for modified concrete preparation. It was a mixture of aggregate - quartz sand, cement, carbonate meal and modifying additives. It consisted 20% of Portland cement and ~80% of quartz sand with maximum grain size of 4 mm. Bulk density of mixture was 1700 kg/m3 ± 5%. The ratio of deionized water added to the dry mixture was, according to the manufacturer's recommendations, in all composites 0.12. The mortar fills the requirements of the PN-EN 13813 CT-C20-F5 standard. The photocatalyst used to create composite was always accounted to 10% wt/wt of cement (2% whole mixture) in the composite.

2.2 Photocatalytic activity tests

In Fig. 1 the scheme of photocatalytic reactor used in this study was shown.

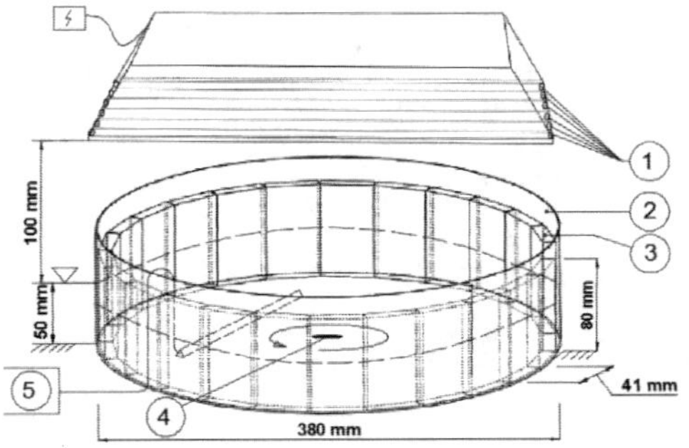

Fig. 1. Scheme of photocatalytic reactor (1. 6x 15 W UVA lamps; 2. steel reactor; 3. concrete plates; 4. magnetic stirrer 5. oxygen concentrator).

The reactor was simple model of the secondary settler, and was stainless steel tank, volume 10 L. A source of UV-Vis light was Magic 836 set (Efbe-Schott GmbH) equipped with six UVA - Vis (UV300 < λ < VIS600) Cleo ISOLde lamps (Phillips), each 15 W. Lamps were 100 mm above the water, parallel to the water surface.

Concrete slabs were laid in a reactor along a steel wall, forming a regular polygon. A dimensions of all concrete plates were the same – 41 x 80 x 10 mm. A number of plates in each reaction was set to 25. The total active surface of photocatalytic underwater area was 512.2 cm^2. In the middle of the bottom of the reactor was placed a magnetic stirrer, ensuring a mixing of the solution. In reactions which using additional oxygenation, it was provided by an oxygen concentrator (Everflo, Phillips) by continuously dosing oxygen at concentration up to 95%, through diffuser with a volume flow of 300 L/h. Due to the necessity of gas injection, in all processes using additional O_2 dosing, prior to the addition of solution to reactor were carried out 1 h photolysis of surfactant molecules by 150 W medium pressure UV lamp. The effect of this process was to reduce foaming of solution to level which allowed to begin the reaction.

All photocatalytic activity studies were made for pure concrete, concrete with 10% wt/wt TiO_2/N and the concrete with 10% wt/wt TiO_2-P25 composite. Photoactivity was examined in the oxidation reactions of the NaDBS surfactant at initial concentrations of 20 mg/L and at unmodified pH. After adding the surfactant solution to the reactor, for 0.5 h at darkness was achieving the equilibrium concentration of organic compound by dynamic adsorption/desorption processes on concrete plates surface. Concentration of total organic carbon (TOC) and UV200-300 absorbance (ABS) value after this process was treated as initial (TOC_0 and ABS_0).

After that, the reactions were started. Studied processes were carried out for 6 or 18 h. The initial TOC concentration of 20 mg/L NaDBS solution was 12 mg/L. The initial concentration of TOC of UV pretreatment 20 mg/L NaDBS solution was 11.8 mg/L.

The efficiency of the each process was determined as the ratio of the TOC concentration and ABS value in the reaction time to their initial values (TOC_t/TOC_0 and ABS_t/ABS_0). UV absorbance spectrum of the reaction mixture samples were determined by UV-Vis spectroscopy (Spectroquant Pharo 300, Merck Millipore). The TOC concentrations were measured on Multi N/C 2100 S (Analytik Jena).

3 Results and discussion

3.1 Decompositon of NaDBS molecules

The oxidation efficiency of material composite consisting nitrogen modified TiO_2 compared to composite with reference TiO_2-P25, pure concrete and control test without any building materials under UV irradiation were presented in Fig. 2. All tested processes lasted 18 h and were oxidized only with natural air circulation. The pH and conductivity changes in time in Table 2 were presented.

Table 2. pH and conductivity changes after 6 and 18 h of UV light irradiation.

Reaction time [h]	Control tests	Con-crete	Con-crete +TiO_2-P25	Con-crete +TiO_2/N
		pH		
0	6.2	10.4	11	10.6
6	6.38	11.05	11.49	10.82
18	6.35	11.01	11.6	10.52
		Conductivity [μS/cm]		
0	16	100	216	147
6	19	441	916	550
18	21	346	733	527

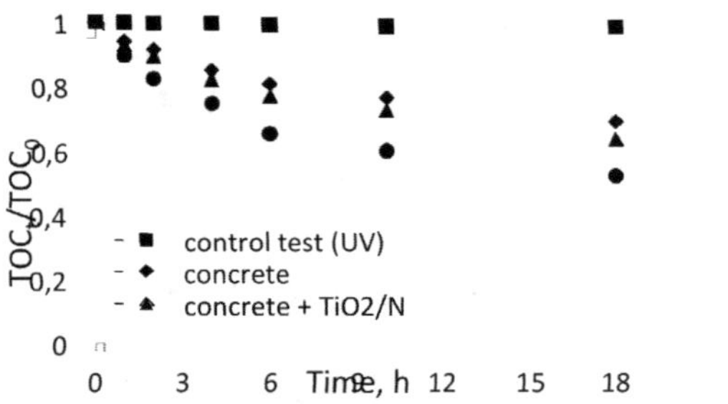

Fig. 2. Changing of TOC content as a function of time during NaDBS 20 mg/L solutions photodecomposition.

In photolysis control test had not been noticed any oxidation reactions. In all others reactions, significant photooxidation processes have taken place. As it can be seen, the most active composite in this condition is the concrete with commercial TiO_2-P25. The mineralization degree after 18 h of reaction with this building material reached 48% of initial TOC. The TOC removal after 18 h in reaction with TiO_2/N/concrete composite was 37%, and 31% in pure concrete. The significantly higher TiO_2-P25/concrete decomposition potential than in the other tested water purification methods was also evidenced by much more intense changes in oxidized solution conductivity (Table 2). The high growth of conductivity testifies to the appearance of a large amount of ionic reaction intermediates and therefore, indirectly, higher degree of surfactant molecule fragmentation. After 6 h of reaction the conductivity value reached 0.91 mS/cm, while in analogous time of reaction with concrete and TiO_2/N catalyst, cond. was much lower and stopped at 0.55 mS/cm.

One of the normal properties of new concrete in is increasing the pH value of water. It is mostly due to calcium hydroxide, which is a normally by-product of cement hydration. In Table 2 was presented, how the value of pH were changing in time of studied reactions. Besides of control test, all of these processes were carried out at very alkaline pH, what could have had a significant impact on the course of the reactions. When the concentration of OH- ions is higher, the ability to consequently trapping holes increasing, resulting in more photogenerated •OH radicals. More effective semiconductor in the surface production of OH radicals than TiO_2/N, as proved Bubacz et al. [19] is TiO2-P25. An interesting phenomenon is, how effective in oxidation process is pure concrete. This also may be related to the ability to surface-forming OH radicals but it requires further, more in-depth research.

At alkaline pH, also the ability to oxidize the anionic surfactant directly on the surface is reduced by lower degree of adsorption because of repulsive effect of same charges (catalyst surface – anionic part of surfactant molecule). This is due to the high negative electrokinetic potential of these two catalysts, with a higher value on TiO_2/N under these pH conditions (Fig. 3). This relationship, in addition to a lower surface hydroxylation capability, have had a decisive effect on TiO_2/N lower efficiency in TOC oxidation. The efficiency of photodecomposition processes can be verified by studying the change in UV absorbance intensity in time.

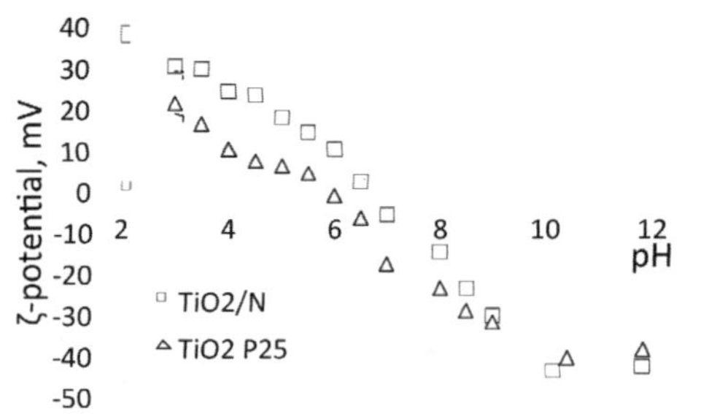

Fig. 3. Relationship between ζ-potential and solution pH value for TiO2/N and TiO2-P25.

Changes in UV absorbance spectra at 223 nm in NaDBS 20 mg/L solutions photodecomposition by various methods in Fig. 4 are presented.

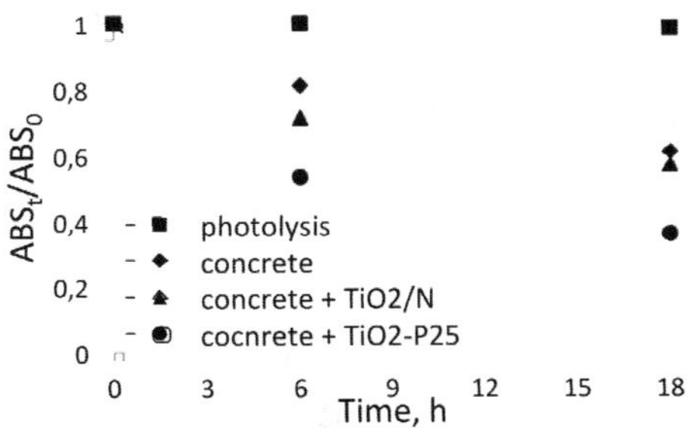

Fig. 4. Changes in UV absorbance spectra at 223 nm in NaDBS 20 mg/L solutions photodecomposition by various methods.

The UV-Vis absorbance spectra of NaDBS during light irradiation showed reveal strong absorption band, centered at and 223 nm, and a second but weak band at ~260 nm. It was found, that they are caused by the π - π electronic transitions of C=C bonds in the aromatic ring of NaDBS. Intensive reduction of the absorbance in these areas proves the decomposition of the organic compound, which also results in a reduction in the propensity to produce foam by the solution. Changes in absorbance at the 223 nm wavelength allow us to suppose that there are changes in the aromatic rings of the surfactant particles. From the analysis of changes in the absorption spectrum of the NaDBS solution during its photooxidation, it follows that the reduction of the absorption band intensity from the maximum at 223 nm is accompanied by an increase in absorption in the range $\lambda \leq 215$ nm. This indicates the formation of reaction products whose absorption spectra are within the analyzed spectral range and overlap with the NaDBS absorption spectrum.

As in the case of TOC mineralization, in reducing the intensity of absorbance also the most effective composite was the one formed from the addition of P25 to concrete and after 18 h resulting in 64% reduction of absorbance at 223 nm wavelength. In concrete with TiO_2/N reaction the reduction of absorbance at this wavelength was 43%, and 39% in pure concrete respectively.

4 Conclusions

Two different types of photocatalysts were studied, commercial TiO_2-P25 and nitrogen modified TiO_2 which were used for two types of modified concretes preparation. It was proved that use of photocatalytic concrete can be successfully applied to reduction of concentration and mineralization of anionic surfactants compounds. Higher activity under UVA irradiation due to its properties was observed with TiO_2-P25 reactions than TiO_2/N catalyst. The mineralization of anionic surfactant NaDBS results achieved by them after 18 h of reactions were respectively: 48% and 37%.

Project was founded by the National Science Centre allocated on the basis of a decision number DEC-2011/01/D/ST5/03467.

5 References

1. S. Talmage, *Environmental and human safety of major surfactants* (Lewis Publisher, 1998)
2. M.I. Rosen, *Surfactants and Interfacial Phenomena* (New York: Wiley, 2004)
3. J. Falbe, *Surfactants in Consumer Products, Theory Technology and Application*, (Springer Verlag, Berlin, 1988)
4. Global Report on the Surfactants Market, Anti-Corros Method M, 59 (2012)
5. Y. Guang-Guo, Environ. Int. **32**, 417 (2006)
6. J.A. Field, T. Poiger, T.M. Field, W. Giger, Water Res. **29**, 1301 (1995)
7. J. Perkowski, L. Kos, Sci. Eng. **18**, 73 (1996)
8. P. Koelbener, U. Baumann, T. Leisinger, Environ. Toxicol. Chem. **14**, 571 (1995)
9. L. Cavalli, A. Gellera, A. Landone, Environ. Toxicol. Chem. **12**, 1777 (1993)
10. M. Perez, L. Romero, D. Sales, J.M. Quiroga, Tenside, Surfactants, Deterg. **33**, 473 (1996)
11. R.I. Bickley, F.S. Stone, J. Catal. **31**, 389 (1973)
12. Z. Zainal, L.H. Hui, M.Z. Hussein, Y.H. Taufiq-Yap, A.H. Abdullah, I. Ramli, J. Hazard. Mater. B. **125**, 113 (2005)
13. Q. Xiao, Z. Si, J. Zhang, C. Xiao, X. Tan, J. Hazard. Mater. **150**, 62 (2008)
14. K. Ishibashi, A. Fujishima, T. Watanabe, K. Hashimoto, J. Photochem. Photobiol. A. **134**, 139 (2000)
15. K. Zając, M. Janus, K. Kuźmiński, A. W. Morawski, Przem. Chem. **95**, 2222 (2016)
16. C. Jun C., P. Chi-Sun, J. Environ. Manage. **90** (11), 3436 (2009)
17. A. Nazari, S. Riahi, Cem. Wapno Beton, **16**, 167 (2011)
18. A. Fachini, I. Joekes, Coll. Surf. A: Physicochem. Eng. Aspects **201**, 151 (2002)
19. K. Bubacz, B. Tryba. A.W. Morawski, Mater. Res. Bull. **47**, 3697 (2012)

Photocatalytic decomposition of surfactants on nitrogen modified TiO_2

Kamil Kuźmiński[1], *Antoni W.* Morawski[2], and *Magdalena* Janus[1,*]

[1]West Pomeranian University of Technology, Faculty of Civil Engineering and Architecture, Department of Sanitary Engineering, al. Piastów 45, 70-311 Szczecin, Poland
[2]West Pomeranian University of Technology, Faculty of Chemical Technology and Engineering, Institute of Inorganic Chemical Technology and Environment Engineering, ul. Pułaskiego 10, 70-322 Szczecin, Poland

Abstract. In these studies advanced oxidation processes such as: photolysis, ozonation and photocatalysis for anionic and cationic surfactants decomposition were used. Nitrogen modified titanium dioxide and commercial TiO_2-P25 were used for photocatalytic tests. UV-C lamp and different dose of ozone: 186, 383, 478 and 563 mg/(dm^3·h) were used. The optimal system for anionic and cationic surfactants decomposition was connection of ozonation with UV-C irradiation.

1 Introduction

Surface active compounds used in households and numerous industrial sectors are one of the pollutants of surface water and groundwater. They constitute a serious water pollution, resulting in ecological threats to plants and animals, caused by the relatively long life of surfactants in water and low susceptibility to natural degradation. Consequently, the use of advanced oxidation methods based on, for example, the photocatalytic action of titanium dioxide particles and the oxidizing properties of ozone, appear to be a good and simple method for waste water purification. These processes are an effective alternative to conventional methods.

The primary and most valuable feature of photocatalysis is the ability to obtain complete mineralization of pollutants. Only effective access to the process requires access to light which may excite a photocatalyst. It has been proven that photocatalytic processes are ideal for the purification of contaminated water with organic compounds such as anionic surfactants sulphonated as sodium lauryl sulfate [1], sodium dodecylbenzenesulfonate [2] or sodium naphthalenesulfonate [3]. These processes have also been successfully used for nonionic decomposition [4] and cationic surfactants [5].

In heterogeneous photocatalysis using TiO_2, the most effective oxidizing compounds are hydroxyl radicals. In aqueous environments they are generated by water oxidation, and are adsorbed on the surface of the photocatalyst. The oxidation agent of catalyst is positively charged electron holes (h +) formed in the TiO_2 valence band a result of its excitation [6].

$$TiO_2 + hv \rightarrow h^+ + e^- \quad (1)$$

$$h^+ + H_2O \rightarrow OH^{\bullet} + H^+ \quad (2)$$

In aerated systems, electrons (e^-) transferred to the conduction band can reduce dissolved oxygen to superoxide anion and continue to peroxide anion [7]:

$$e^- + O_2 \rightarrow O_2^{\bullet -} \quad (3)$$

$$e^- + O_2^{\bullet -} \rightarrow O_2^{2-} \quad (4)$$

It has been proven that photocatalysis can be an effective way of mineralizing surfactants, but combining it with the ozonation process can definitely accelerate the oxidation of pollutants. The general mechanism of oxidation of ozone in the aquatic environment can be divided into direct cycloaddition to multiple bonds (ozonolysis) [8], having a significant effect on the removal but not mineralization, and indirect mineralization by O_3 decomposition products [9]. Ozone in the aquatic environment decomposes as a result of radical chain reactions, whose main products are OH radicals and H_2O_2 molecules.

The main aim of the study was to compare the photocatalytic efficiency of oxidation of anionic and cationic surfactants using nitrogen modified titanium dioxide (TiO_2/N) obtained from the Institute of Inorganic Chemical Technology and Environment Engineering of the West Pomeranian University of Technology with commercially available TiO_2-P25. In addition, the effect of changes in the ozonation process parameters in combination with UV photolysis and photocatalysis was determined.

2 Experimental

2.1 Materials

Sodium dodecylobenzenosulfonate and benzethonium chloride with concentration of 20 mg/dm^3 ($5.75 \cdot 10^{-5}$ mol/dm^3 NaDBS, 11.9 mg/dm^3 pH 5.4;

* Corresponding author: mjanus@zut.edu.pl

$4.46 \cdot 10^{-5}$ mol/dm^3 BtCl, TOC 16.2 mg/dm^3, pH 5.2) were used as model surfactants.

Commercial TiO_2 P25 was used as reference material with crystal size 26 nm and BET surface area 56 m^2/g, 75% of anatase and 25% of rutile phase [11]. At West Pomeranian University of Technology there is a possibility to produce nitrogen modified photocatalyst with high photocatalytic activity and high adsorption capacity. Nitrogen modified TiO_2 was obtained from Institute of Inorganic Chemical Technology and Environment Engineering. It was choose as a second photocatalyst because there was no data about its activity during surfactants decomposition Moreover this is photocatalysts with different characteristic than commercial TiO_2 P25. It was prepared at E746 AutoLab (HEL Ltd) installation. TiO_2-N average particle size amounted to 190 nm, its BET surface area is 345 m^2/g, amorphous phase 67%, anatase phase 30%, rutile phase [10].

Zeta potential of TiO_2/N and TiO_2 P25 was determined for pH 3 and 6, because the reactions of photocatalytic decomposition of surfactants were conducted in these pH ranges. In Fig. 1 the relationship between ζ-potential and solution pH value for tested photocatalysts was presented.

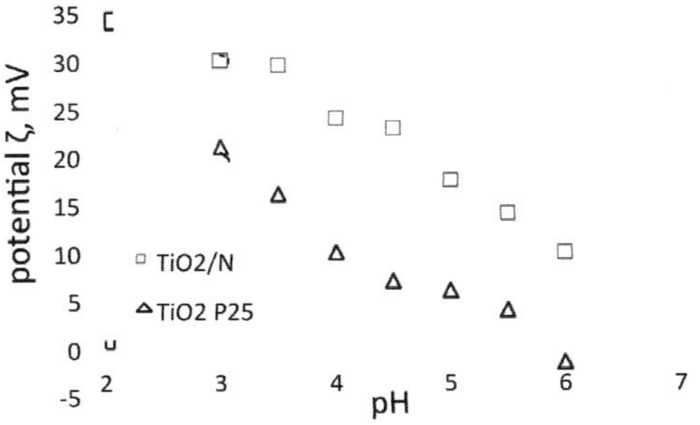

Fig. 1. Relationship between ζ-potential and solution pH value for TiO_2/N and P25.

2.2 Advanced Oxidation Processes

In Fig. 2 the installation for advanced oxidation processes is presented. The solution in a stainless steel reactor with an active capacity of 5 dm^3 was in the continuous flow provided by a peristaltic pump. The reactor was equipped with a low-pressure UV-C (25W) mercury lamp emitting almost all energy at 253.7 nm (TMA, Poland). Ozone was produced in the O3PRO DRP-30.7VW generator (Prozonex, Poland), powered by an oxygen concentrator (Everflo, Phillips) and pressed into the bottom of the UV-C reactor and then distributed evenly through the diffuser. The ozone concentration was measured in real time on the conduit connecting the O_3 generator to the UV-C reactor via a photometric gas ozone analyzer (BMP 964, BMT Messtechnik GMBH, Germany) with a measurement range of 0-300 g/Nm^3. The concentration of total organic carbon (TOC) was determined using the Multi N/C 2100S analyzer (Analytik Jena AG, Germany).

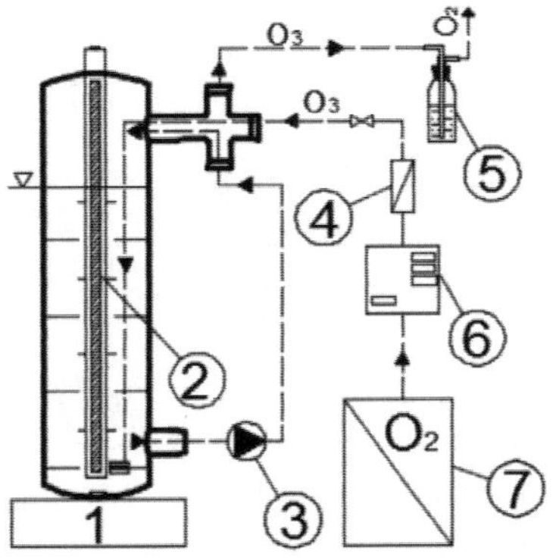

Fig. 2. Scheme of installation for AOPs (1 - magnetic stirrer, 2 - UV-C lamp, 3 - peristaltic pump, 4 - ozone analyzer, 5 - KI scrubber, 6 - ozone generator, 7 - oxygen concentrator).

The initial concentration of surfactants in aqueous solutions was 20 mg/dm^3 ($5.75 \cdot 10^{-5}$ mol/dm^3 NaDBS; $4.46 \cdot 10^{-5}$ mol/dm^3 BtCl). Prior to photocatalysis (TiO2 + UV) and photocatalytic ozonation (TiO2 + UV + O3) processes, adsorption was carried out in the dark for 30 minutes (to obtain an adsorption equilibrium). The concentration of the total organic carbon (TOC) after this step was treated as the initial concentration (TOC_0). UV-C (UV) photolysis, ozone (O_3) photolysis, UV-C photolysis with simultaneous ozone (UV + O3) was also performed to compare the effectiveness of the investigated processes. During the photocatalytic reaction, oxygen was injected into the reactor (100 dm^3/h).

3 Results and discussion

3.1 Ozonation and UV light irradiation

For decomposition of anionic and cationic surfactants the ozone and UV-C light irradiation were used. As it can be observed in Table 1 under UV light irradiation the increasing of ozone dose (from 186 to 563 mg/(dm^3·h)) increased the surfactants decomposition. Ozone alone removed only 9% of NaDBS and 20% of BtCl. The initial pH of reaction solution is acidic and in this condition mainly ozonolization occur, but the oxidation of BtCl is faster than NaDBS, because O_3 reacts with chlorine anions according to equations 5-6:

$$H^+ + O_3 + Cl^- \leftrightarrow HO_3Cl \quad (5)$$

$$HO_3Cl \rightarrow HOCl + O_2 \quad (6)$$

As a result hypochlorous acid is formed, widely used as a bleaching agent and disinfectant due to its high oxidizing potential. Another reason could be the electrophilic behavior of the ozone molecule relative to the carbon atoms in the α position to the oxygen atoms (at pH <6). As a result of the transfer of the hydrogen atom to the O_3 molecule, a carbonic cation is formed, stabilized by resonance with the oxygen atom.

Table 1. Removal of TOC content in NaDBS and BtCl solution during advanced oxidation processes after 1 h, values of dosed ozone are shown in brackets (mg/(dm³·h)).

	% removal of NaDBS	**% removal of BtCl**
UV-C	11	14
O_3 (478)	9	20
UV-C + O_3 (186)	45	47
UV-C + O_3 (383)	86	82
UV-C + O_3 (478)	96	84
UV-C + O_3 (563)	97	88

To the cation, the previously formed HO_3^- may be attached to form a HO_2^- fast reaction with H^+ followed by removal of one oxygen atom, which is the product of H_2O_2. This follows, in line with the course of the following reaction, forms OH radicals with the ozone [13]:

$$2O_3 + H_2O_2 \rightarrow 2OH^{\bullet} + 3O_2 \tag{7}$$

Under pH higher than 7 the OH• radicals are produced (equations 8-10) [9]:

$$O_3 + OH^- \rightarrow O_2^{\bullet-} + HO_2^{\bullet} \tag{8}$$

$$O_3 + HO^{\bullet\cdot} \rightarrow O_2 + HO_2^{\bullet} \leftrightarrow O_2^{\bullet-} + H^+ \tag{9}$$

$$O_3 + HO_2^{\bullet} \rightleftharpoons 2O_2 + OH^{\bullet} \tag{10}$$

In Table 2 the % removal of surfactants with combined process of UV-C irradiation, ozonation and photocatalysis was presented. In Fig. 2 a) and b) the kinetics of surfactants decomposition under UV-C irradiation is presented.

Table 2. Photocatalytic decomposition of NaDBS and BtCl after 1 h of irradiation, in brackets are shown values of dosed ozone (mg/(dm³·h))T.

	% removal of NaDBS	**% removal of BtCl**
UV-C-P25	37	52
UV-C-TiO_2/N	20	55
O_3 (383) + UV-C-P25	89	83
O_3 (383) + UV-C-TiO_2/N	78	81

The charge on the TiO_2 surface with the opposite sign to the charge sign of the ionic surfactant will promote the adsorption of this surfactant to the catalyst. In surfactant solutions with acidic pH, higher electrokinetic potential is characterized by TiO_2/N and this is one of the main reasons for higher NaDBS desorption values.

As a result of the exposure process (t = 0) partial photocatalyst excitation occurs, resulting in first photolytic and photocatalytic reactions on the semiconductor surface. As a result of these reactions, the hydrocarbon chains of NaDBS molecules adsorbed on TiO_2, resulting in partial desorption of these compounds, are depicted as increasing the concentration of TOC in the purified solution (Fig. 2). At a further stage of the process, TiO_2-P25 is clearly a more effective semiconductor in the NaDBS oxidation by UV-C photocatalysis. This is due to the fact that TiO_2-P25 is characterized by a significantly higher degree of surface hydroxylation [12], coupled with lower adsorption potential due to the lower zeta potential value, and therefore greater surfactant-free surface and better absorption of radiation. For higher efficiency in oxidation of NaDBS in solution volume by oxidation of water molecules to OH radicals.

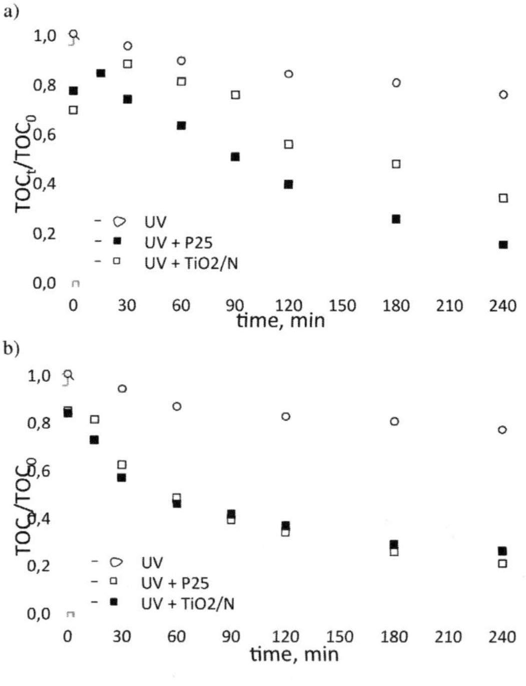

Fig. 3. Changing of TOC content in a) NaDBS and b) BtCl solutions as a function of time during UV-C photolysis and photocatalysis reactions using nitrogen modified TiO_2 (TiO_2-N) and TiO_2 P25 (time 0 – TOC_t/TOC_0 after adsorption equilibrium process).

Photocatalytic surface reactions occur less often because the low-energy UV-C lamp induces a lower proportion of TiO_2/N particles than TiO_2-P25 because most of the active sites of the first are coated with anionic surfactant molecules. As a result, NaDBS oxidation through the electron hole in the valence band may lose its importance, making TiO_2-P25 a better photocatalyst under given conditions.

3 Conclusions

As a result of the studies, it was found that the anionic NaDBS surfactant was more susceptible to mineralization using advanced oxidation methods than the cationic BtCl surfactant. The best condition for anionic and cationic surfactants decomposition is: UV-C light and ozone dosed in 478 $mg/dm^3 \cdot h$. Under these conditions decomposition of NaDBS after 4 h of irradiation amounted to 96% and decomposition of BtCl amounted to 84%.

References

1. O. Horváth, R. Huszánk, Photochem. Photobiol. Sci. **2**, 960 (2003)
2. K. Kuźmiński, J. Grzechulska-Damszel, K. Zając, A.W. Morawski, M. Janus, Przem. Chem. **95**, 2210 (2016)
3. P. Avetta, A. Bianco Prevot, D. Fabbri, E. Montoneri, L. Tomasso, Chem. Engin. J. **197**, 193 (2012)
4. J. Perkowski, A. Bulska, W.K. Jóźwiak, Environ. Prot. Eng. **31**, 61 (2005)
5. H. Hidaka, S. Yamada, S. Suenaga, Z. Jincai, J. Mol. Catal., (1990)
6. M.R. Hoffmann, S.T. Martin, W. Choi, D.W. Bahnemann, Chem. Rev. **95**, 69 (1995)
7. Z. Zsilak, O. Fonagy, Environ. Sci. Pollut. Res. **21**, 11126 (2014)
8. H.R. Einsenhauer, Water Res. **5**, 467 (1971)
9. B. Kasprzyk-Hordern, M. Ziółek, J. Nawrocki, Appl. Catal. B. **46**, 639 (2003)
10. M.D. Hernandez-Alonso, J.M. Coronado, A.J. Maira, J. Soria, V. Loddo, Appl. Catal. B. **39**, 257, (2002)
11. A.V. Levanov, I.V. Kuskov, E.E. Antipenko, V.V. Lunin, Russ. J. Phys. Chem. A. **86**, 757 (2012)
12. D.H. Giamalva, D.F. Church, W.A. Pryor, J. Amer. Chem. Soc. 108, 7678 (1986)
13. R. Munter, Proc. Eston. Acad. Sci. Chem. **50**, 59 (2001)

An evaluation of the adsorption process in removing organic substances in a pilot flow-type surface water treatment system

Małgorzata Wolska[1,2,*], *Marek* Mołczan[1,2], *Sławomir* Szerzyna[1], *Wojciech* Adamski[1], *Jacek* Wiśniewski[1], *Zbigniew* Ferenc[1], *Aleksandra* Sambor[1], and *Małgorzata* Żółtowska[2]

[1]The Department of Water and Sewage Treatment Technology, Wrocław University of Science and Technology, Wybrzeże Wyspiańskiego 27 Street, 50-370 Wrocław, Poland
[2] Municipal Water and Sewerage Company in Wrocław, Na Grobli 14/16 Street, 50-421 Wrocław, Poland

Abstract: This article presents the results of the adsorption process effectiveness in treating surface water, especially in removing organic substances. The effectiveness of the adsorption system was evaluated at different levels of adsorption capacity exhaustion of the activated carbon bed, which was possible due to replacement of the beds during the study period. Studies have shown that among the removed substances, chlorinated disinfection byproduct precursors dominated, and during the period preceding bed replacement, the biological activity of microorganism populating the activated carbon ensured a reduction in not only organic but also non-organic food substrates. In such a adsorption bed populated with microorganism a nitrification process took place, indicating a high degree of process stability. Replacing the adsorption beds provided a significant increase in the effectiveness of removing organic substances, especially those absorbing UV light, therefore removing chlorinated organic disinfection by-product precursors.

1 Introduction

The increase in contamination of source surface waters due to their contamination with household and industrial wastewater and atmospheric precipitation has caused that water treatment systems more often make use of adsorption processes [1,2]. This process ensures the removal of organic substances of medium and low molecular mass [3-7], whose elimination is not possible during the coagulation process, which is commonly used for removing organic substances from surface waters before the adsorption process. Additionally, as many studies show [8,9], adsorption of activated carbon allows for a elimination of various types of microcontaminants. More frequently, different groups of microcontaminants are detected in waters, and within these groups a constantly increasing number of chemical compounds being introduced into the environment. The adsorption process effectively removes some pharmaceuticals [10], pesticides [11], phenols [12] and non-organic microcontaminants such as heavy metals [13]. This is one of the reasons of the increasing use of activated carbon in surface water treatment technology.

Surface water treatment technology also commonly makes use of that fact that adsorption beds are easily populated with microorganisms, which form a biofilm on the bed surface [14,15]. Biologically activated beds yield an increase in organic substance removal due to biodegradation by microorganisms. [16]. An increase in the effectiveness of biologically active beds in removing organic substances is also achieved by preceding adsorption process with ozonation [8] which yields a change in the structure of organic substances in water towards one more susceptible to adsorption and biodegradation.

The effectiveness of adsorption depends on many factors and decreases along with increasing bed operating time, and therefore the exhausting of adsorbent potential. . Generally, in technical-scale systems, the need for bed replacement/regeneration is decided by, apart from worsening of filtrate quality, by the iodine number value

The study presented in this paper was conducted at a pilot scale in a flow-type system supplied by a surface water treatment plant, which allowed for an evaluation of the changes of the adsorption process efficiency with changes in raw water quality in conditions of continuous and long-term operation.

2 Methods of study

Studies of adsorption process effectiveness in the surface water treatment system were conducted in a pilot-scale flow-type system with a throughput of 3 m^3/h consisting of coagulation, sedimentation, filtration, adsorption, disinfection and pH correction. The treatment system was

* Corresponding author: wolska@pwr.edu.pl

https://doi.org/10.1051/e3sconf/20185900018

supplied with the same water as the water treatment plant treating water for human consumption. The adsorption process was conducted in the system in pressure filters with a bed with granulated activated carbon, whose properties are shown in Table 1.

Table 1. Parameters of fresh WG-12 activated carbon.

No	Parameter	Unit	Value
1.	Bulk density	g/dm^3	470
2	Volatile content	%	1.50
3	Ash Content	%	11.6
4.	Specific Surface Area	m^2/g	968
5.	Iodine Number *	mg/g	1014
6.	CTC –adsorption CCl_4	%	62.30
7.	<0,5 mm subfraction content	%	0.00
8.	Mechanical strength	%	97.30

*value specified by manufacturer of fresh activated carbon

The beds were filled in June 2016, and were replaced in September 2016, which makes it possible to evaluate the effectiveness of the adsorption process in different phases of bed adsorption capacity exhaustion. During the bed replacement, an analysis of the iodine number of the bed being removed was performed, which yielded a value of 660 mg/g, which indicated that the adsorbing potential of this bed had not yet been exhausted. This same value for the fresh bed was 1040 mg/g.

Despite operation the system at a constant throughput, the adsorption filters were operated at varying operating parameters, which resulted from the necessity of continuous system operation (Table 2).

Table 2. Adsorption bed operating parameters.

Parameter	Unit	Value
Filtration area	m^2	0.59
Bed height	m	1.50
Filtration Rate	m/h	4.71-5.15
Water-Activated Carbon Contact Time	min	17.40-19.0

The study was conducted during the time period from May 2016 to December 2016, with samples being taken weekly before and after the adsorption process. The long time period of the study, and therefore the large variation in raw water, allowed for assessment of the impact of supply water quality on the adsorption process effectiveness.

In all water samples, the following parameters were measured: T, pH, specific conductivity, color, turbidity, UV absorbance UV_{254} and UV_{272}, total (TOC) and dissolved (DOC) organic carbon concentration, non-organic nutrient concentrations (ammonia ion, nitrites, nitrates, phosphates) and concentrations of chlorides and sulfates. In all pre- and post-adsorption water samples, the total psychrophillic and mesophillic bacteria numbers were also determined.

Based on UV_{254} absorbance values, and DOC concentrations the specific UV absorbance values (SUV-a) was calculated.

All analysis of water quality indicators were conducted with current Polish standards, according to accredited laboratory procedures.

3 Results and discussion

The source water quality during the study period was characterized by a very large variability in composition, which consequently resulted in a decrease in quality of water undergoing adsorption (Table 3). The variability in composition was however lower in water before adsorption than that found for raw water. This means that the coagulation process used before adsorption, along with sedimentation, ensured an effective elimination of organic substances and significantly reduced the variability of water quality reaching the adsorption beds [17].

The greatest variability in water reaching the adsorption filters, apart from temperature, was found for indicators relating to organic substance content (TOC, DOC, UV_{254} and UV_{272} absorbances and color). Among the organic substances in water supplying adsorption beds, the dissolved fraction dominated (from 96.9% to 99.1%), which testifies to the large susceptibility of organic substances to removal by adsorption. Furthermore, the calculated SUV-a values in the range of 1.65-2.20 m^2/g testify to the limited susceptibility of organic substances to removal during this process [18].

The presence of non-organic nutrient substrates in water supplying the beds (Table 3) and microorganisms which may allow for biofilm development (Fig. 1), along with the long adsorption bed operating time suggest the conclusion that, apart from adsorption, biodegradation also took place within the beds.

The effectiveness of the adsorption process is confirmed by the elimination of organic substances that was found, which during the study was in the ranges of 11.5-59.4% and 10.8-61.2% for TOC and DOC respectively (Fig. 2).

A significant increase in organic substance removal effectiveness was found after bed replacement, which confirms the partial exhaustion of bed exhaustion potential before the replacement. On the other hand, the successive decrease in organic substance removal effectiveness in the first two months after bed replacement may be explained by bed break-in and the development of biofilm on bed grains. After this time period, a certain equilibrium developed between adsorption and biodegradation.

Table 3. Ranges of water quality indicators before and after absorption.

Indicator	Unit	Before Adsorption	After Adsorption
T	°C	6.70-23.30	7.70-23.50
pH		6.73-7.72	6.65-7.60
Basicity	val/m^3	2.00-3.50	2.00-3.50
Mt	NTU	0.10-0.80	0.00-0.60
B	g/m^3	4.60-7.60	0.40-4.30
UV_{254}	M	5.94-8.38	0.44-5.58
UV_{272}	M	4.91-6.87	0.36-4.53
TOC	gC/m^3	2.90-4.60	1.82-4.77
DOC	gC/m^3	2.86-4.33	1.63-4.23
NO_3^-	gNO_3^-/m^3	2.68-21.66	2.86-21.75
NO_2^-	gNO_2^-/m^3	0.00-0.03	0.00-0.03
NH_4^+	gNH_4^+/m^3	0.00-0.07	0.00-0.07
Cl^-	gCl^-/m^3	24.30-55.90	24.00-55.90
SO_4^{-2}	gSO_4^{-2}/m^3	63.80-129.90	63.80-130.20
PO_4^{-3}	gPO_4^-3/m^3	0.00-0.04	0.00-0.03
SUVA	m^2/g	1.65-2.29	0.46-1.93

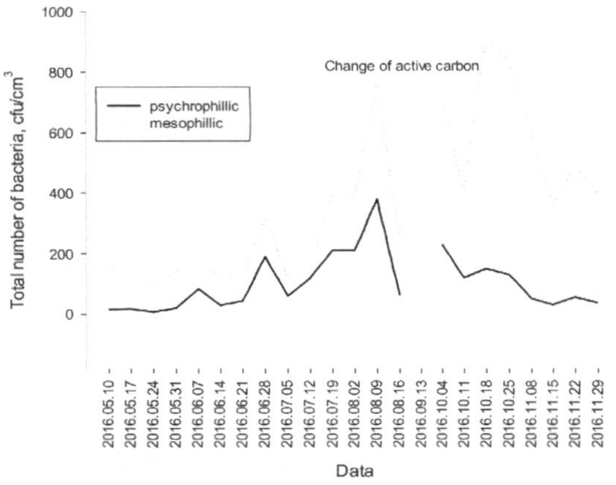

Fig. 1. Variation in numbers of psychrophillic and mesophillic bacteria in water supplying adsorption beds.

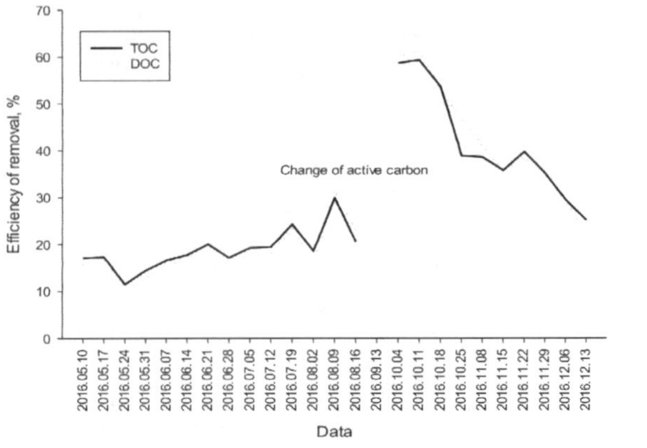

Fig. 2. Total and dissolved organic carbon removal effectiveness during adsorption.

The following nearly constant effectiveness corresponds to that obtained in bends before replacement. It is likely that the reduction in adsorption effectiveness was accompanied by an increase of the degree of biodegradation, which ensured a continued constant process effectiveness for a long time period. This points to the fact that exhaustion of the beds did not take place in any of the beds, which is also confirmed by the iodine number of the replaced bed.

The obtained effectiveness in reducing TOC and DOC concentrations before bed replacement testifies therefore to the simultaneous nature of adsorption and biodegradation. The biological effectiveness of the adsorption beds before replacement is also testified to by the simultaneous removal in ammonia ion content and phosphate concentrations, i.e. non-organic food substrates for heterotrophic bacteria [19]. The effectiveness in removing ammonia ions was in the range of 7.1-75.0% and was accompanied by an increase in nitrates (found in the time period before bed replacement), which may point to a process of nitrification in the absorption bed. This hypothesis is confirmed by the presence of dissolved in oxygen in water entering the beds, and the decrease in its concentration in the range 1.12-1.72 gO_2/m^3 during flow through the bed. The nitrification process was also aided by the small variation in water flow through the system, yielding stable bed operating conditions.

After bed replacement, however, a significant increase in phosphate concentrations was found, as phosphates were probably flushed out from fresh activated carbon. With respect to non-organic nitrogen compounds, no general trend was found, which may point to their simultaneous use by microorganisms and their release from activated carbon. Such a lack of clear decrease or increase is the result of these mechanisms. After bed replacement, apart from an increase in phosphate concentration, a significant increase of the total number o mesophillic and psychrophillic bacteria was found, being in the ranges of 799-28944 cfu/cm^3 and 1170-44510 cfu/cm^3 respectively.

It was found that during the study period, the organic substance removal effectiveness was proportional to their content in water reaching the adsorption beds. (Fig. 3). This indicates that the variability in organic substances levels in water entering adsorption beds and the source water contamination levels has a significant impact on effective organic substance elimination.

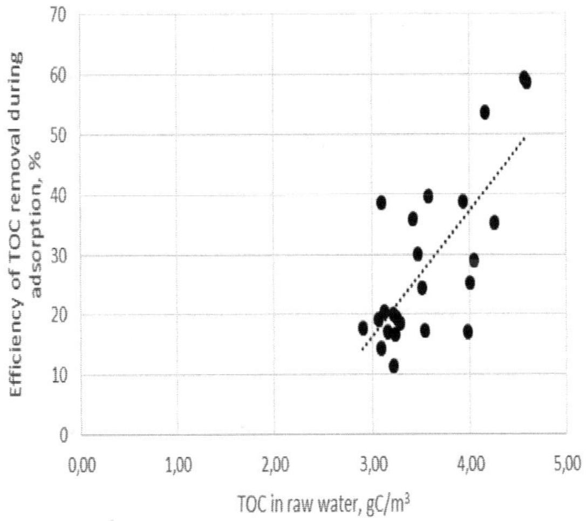

Fig. 3. Effect of organic substance content in raw water on their elimination during the adsorption process.

Among the removed organic substances, UV absorbing substances at wavelength of 254 nm and 272 nm dominated, corresponding to disinfection by-products. The effectiveness in removing dissolved organic carbon was nearly proportional to the effectiveness in decreasing UV absorbance at 254 nm and 272 nm (Fig. 4).

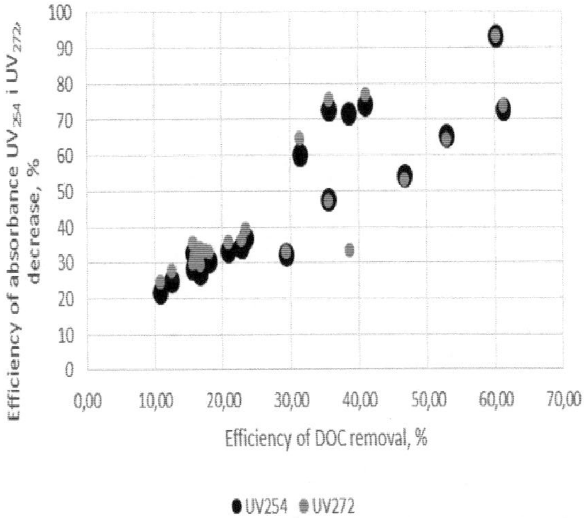

Fig. 4. Dependence of dissolved organic carbon removal and UV absorbance.

According to Korshin et al [20], these precursors exhibit a significantly larger correlation to absorbance at 272 nm than at 254 nm. An effective elimination of chlorinated organic substance precursors regardless of the bed adsorption potential exhaustion ensures a significant limiting of risks associated with disinfection that is used after the adsorption process [17,21].

A consequence of the decrease in organic substance content during adsorption was a decrease in specific UV absorbance in all filtrate samples. This points to the fact the adsorption process removed organic substances of medium molecular mass, and to a lesser degree small organic substances.

Due to the small variability in system throughput, no unambiguous relationship was found between organic substance removal effectiveness and the water flow rate through the bed. Furthermore, no relationship was found between the temperature of water supplying the bed and the adsorption process effectiveness.

4 Conclusions

This study has shown:

1. An effective elimination of organic substances just after filtration bed replacement and achieving an equilibrium between adsorption and biodegradation after two months of operation.
2. Biological activity of filtration beds before replacement ensured the removal of not only organic substances but also non-organic food substrates.
3. In the biologically active adsorption bed nitrification took place, which may testify to biofilm stability on the surface of activated carbon. .
4. Among organic substances, those most effectively removed regardless of adsorption bed exhaustion, were chlorinated organic substance precursors absorbing UV radiation.
5. A slight increase in water-bed contact time has no effect on the effectiveness of the adsorption process.

This publication was made possible by the National Center for Research and Development grant (PBS3/B9/44/2015) "Research on effectiveness of new water treatment technology as a step towards a shift in thinking about water utility sector" (WODTECH), and thanks to involvement of both project consortium members (Wrocław Municipal Waterworks and Drainage Company and Wrocław University of Science and Technology)

References

1. W. Wang, L. Ho, D.M. Lewis, J.D. Brookes, G. Newcombe, Water Res. **17**, 41 (2007)
2. J. Kim, B. Kang, Water Res. **1/2**, 42 (2008)
3. M. Kłos, J. Gumińska, Ochr. Sr. **31**, 3 (2009)

4. H. Ødegaard, S. Østerhus, E. Melin, B. Eikebrokk, Drinking Water Eng. Sci. **1**, 3 (2010)
5. T.D. Lekkas, K.G. Babi, K.M. Koumenides, C. A. Makri, D.T. Lekkas, A.D. Nikolaou, Global Nest J. **11**, 3 (2009)
6. C.O. Lee, K.J. Howe, B.M. Thomson, Water Res. **4**, 46 (2012)
7. J. Reungoat, M. Macova, B.I. Escher, S. Carswell, J.F. Mueller, J. Keller, Water Res. **2**, 44 (2010)
8. F. Hammes, S. Meylan, E. Salhi, O. KÖster, T. Egli, U. von Gunten, Water Res. **7**, 41 (2007).
9. Y. Luo, W. Guo, H.H. Ngo, L.D. Nghiem, F.I. Hai, J. Zhang, X.C. Wang, Sci. Total Environ. **1**, 473 (2014)
10. T. Urase, T. Kikuta, Water Res. **7**, 39 (2005)
11. K. Ignatowicz, J. Hazard. Mater. **1**, 169 (2009)
12. R.I. Yousef, B. El-Eswed, H. Ala'a, Chem. Eng. J. **3**, 171 (2011)
13. C.Y. Cao, J. Qu, W. S. Yan, J.F. Zhu, Z.Y. Wu, W.G. Song, Langmuir. **28**, 9 (2012)
14. H. Wang, L. Ho, D.M. Lewis, J.D. Brookes, G. Newcombe, Water Res. **18**, 41 (2007)
15. E.S. Melin, H. Ødegaard, Water Res. **18**, 34 (2000)
16. A. Pruss, A. Maciołek, I. Lasocka-Gomuła, Ochr. Sr. **31**, 4 (2009)
17. I. Zimoch, B. Kotlarczyk, A. Sołtysik Ochr. Sr. **29**, 3 (2007) http://www.os.not.pl/docs/czasopismo/2007/Zimoch_3-2007.pdf
18. T. Karanfil, M.A. Schlautman, I. Erdogan, AWWA J. **12**, 94 (2002)
19. E.C. Wert, J.J. Neemann, D.J. Rexing, R.E. Zegers, Water Res. **1/2**, 42 (2008)
20. G.V. Korshin, W.W. Wu, M.M. Benjamin, O. Hemingway, Water Res. **13**, 36 (2002)
21. I. Zimoch, Ochr. Sr. **33**, 3 (2011) http://www.os.not.pl/docs/czasopismo/2011/3-2011/Zimoch_3-2011.pdf

Variability of water intake in settlement units adjacent to Wrocław

Wojciech Cieżak[1,*], and *Jan* Cieżak[1]

[1]Wrocław University of Science and Technology, Faculty of Environmental Engineering, Department of Water Supply and Sewerage, ul. Wybrzeże Wyspiańskiego 27, 50-370 Wrocław, Poland

Abstract. Water intake in rural settlements adjoining large cities is variable and depends on many factors. In these places increasingly often a change in the character of water use from rural to the one typical of urban housing estates is observed. This is due to the expansion of housing, especially single-family housing, and migration of people from large cities. As a consequence, the standard of home furnishings improves and water is used not only for domestic purposes, but also for watering gardens and greenery, the filling and operation of swimming pools, etc. Considering the above, the authors present the results of own research relating to the analysis of water intake in villages adjacent to Wrocław.

1 Introduction

The size and distribution of consumer demand for water in the area of a given administrative unit is the basis for dimensioning its water supply and sewerage facilities. It is also a major element in the mathematical modelling of water or sewage flows [1-3].

The estimation of the demand for water on the basis of water distributions is an essential part of planning and designing the development and operation of water supply systems. It is important to determine the consumers' water needs in order to correctly distribute them over the considered area [4-6]. Depending on the aim of the particular study (a general plan, development programming, technical-economic assumptions for water mains construction or extension), a different degree of precision of predicting the future water needs is required. Most often the following three methods are used:
- a method based on global indices of the demand for water for domestic-municipal and production purposes per 1 inhabitant, which is used in general feasibility and conceptual studies. The method uses unit water demand indices which take into account the total water needs per 1 inhabitant. The global indices are determined on the basis of analytical-statistical studies of the water consumption so far, its multi-annual trends and the predicted trends stemming from the country's long-term and targeted socio-economic development assumptions;
- a method based on aggregate indices of water demand for different consumers in a settlement unit area per 1 inhabitant, which can be used in general plans and in technical-economic assumptions for water mains construction and extension for settlement units having a multifunctional character. In this case, the total daily water demand is the sum of the partial needs determined for the different groups of consumers on the basis of the unit aggregated indices and the assumptions concerning the long-term living conditions of the population;
- a method of water demand balancing for different purposes in a settlement unit area, using detailed unit water demand indices. This is the most labour-intensive method, but enabling one to determine the water abstractions more precisely. The balance is drafted on the basis of the development plan's descriptive part and drawings and the detailed water demand indices determined for the particular consumers and the different purposes. The average and maximum daily water demand determined using this method is the sum of the partial demands of the particular categories of consumers.

The method based on aggregate indices and the one based on detailed unit water demand indices are most often used for balancing the current and long-term water needs. The indices and also daily water demand irregularity coefficients are assumed in accordance with the relevant legal acts and the literature on the subject. From among the current legal acts one should mention here the Minister of Infrastructure Order of 14 January 2002 concerning the determination of average water consumption standards, which can serve as the legal basis for drawing a balance of water needs, even though the Order concerns the determination of average water consumption standards for individual water customers in the case when there is no main water meter. The previously issued guidelines for balancing water needs, such as: Technical guidelines for designing the size of a municipal main with regard to water demand; the Order of the Municipal Economy Minister (Bldg. Gaz. No. 16, it. 68, 9.12.1966); Directives for programming water demand and sewage amount in urban settlement units,

* Corresponding author: wojciech.ciezak@pwr.edu.pl

Ministry of Administration, Local Economy and Environmental Protection, 1978; and Directives for programming water demand and sewage amount in urban settlement units, Institute for Spatial and Municipal Economy, 1991; became invalid because of the marked decline in water consumption in the 1990s and the change from the upward trend to a downward trend.

The quoted above Order of 2002 includes average water consumption standards for households, watering home gardens and agricultural crops, services, farms and livestock facilities, servicing motor vehicles, agricultural machinery and workshops, agricultural products and food processing plants, construction works, chemical plant protection and military facilities belonging to the Defence Department and the Ministry of the Interior. However, to this day the directives have not been brought up to date by government institutions.

According to the Order of 2002, the flow of water consumed by households having only a kitchen sink and a toilet (no bathroom and no hot water) amounts to 70-90 liters per capita per day (lpcd) while in the case of households additionally having a bathroom with a local source of hot water, it amounts to as much as 80-100 lpcd. In Poland, households of this type occur mainly in rural areas. The above flow rates are similar to the ones observed over the last decade in rural areas in Poland (Fig. 1 and Fig. 2) [7].

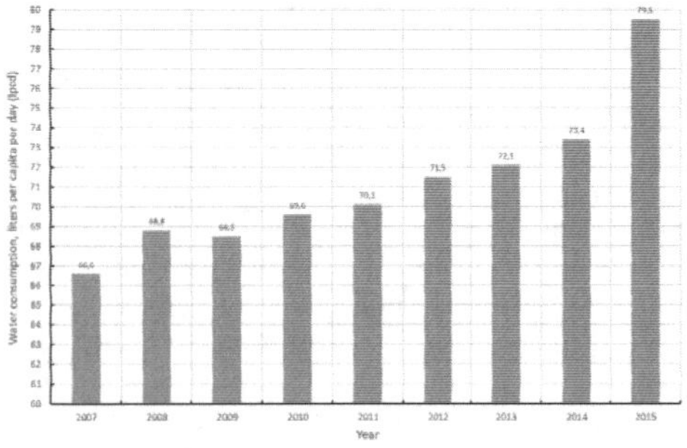

Fig. 1. Consumption of mains water by households in rural areas in the years 2007-2015 [7].

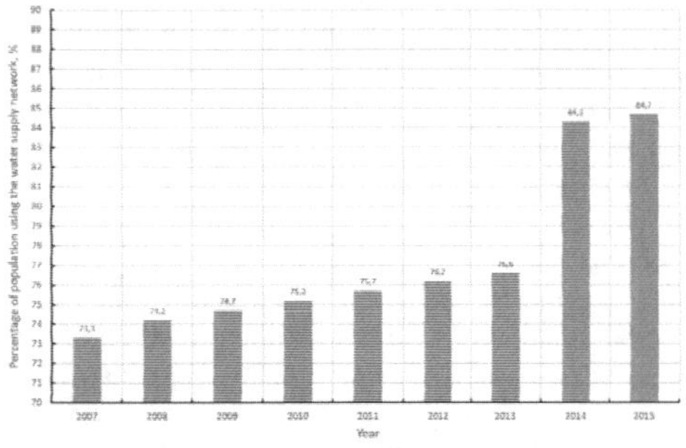

Fig. 2. Percentage of population using the water supply network in rural areas in 2007-2015 [7].

The Order also includes average water consumption standards for, among other things, watering home gardens (including crops, lawns and greenery) – 2.5 L/m^2 per day, and, to a quite a large extent, average water consumption rates for services, livestock facilities and agricultural product and food processing plants.

In the literature on the subject there is relatively little information about water consumption trends in rural settlements, especially the ones adjoining large cities. Since the character of such developments has been changing from rural to the one typical of urban housing estates, it is important to take this fact into account when calculating water needs. Ogiołda and Kozaczek [8] analysed water consumption in the rural water supply systems in the Głogów District. These are areas directly adjacent to Głogów and include the places: Borek, Bytnik, Przedmoście, Zabornia, Wilków, Klucze, Krzekotów and Serby Stare. The calculations and analysis of water consumption were made for data for the years 1995-2011. The mean value of the unit water demand index amounted to 82.2 lpcd for the Borek water supply system and to 68.8 lpcd for the Wilków water supply system while the daily water demand irregularity coefficient amounted to respectively 1.3 and 1.7. Pawełek et al. [9] describe the variation in water consumption in the rural households in 8 villages in the Brzesko District in the years 2000-2012. The analyses showed that the unit water consumption in the studied area ranged from 30.9 to 70.5 lpcd, on average amounting to 46.7 lpcd. Orzechowska [10] studied 16 rural group water mains in the Podlasie province in the years 2006-2010. The author showed that the average water consumption in the analysed area amounted to 86.0 lpcd.

Trade journals most often publish the average unit water demand for water in single-family housing in urban areas [11-13]. The demand usually ranges from 90.0 to 180.0 lpcd. Daily water consumption irregularity coefficients (N_d), expressing the variability of water abstraction by water consumers, are reported much less frequently. According the directives (1978, 1991), the values of N_d for single-family housing are assumed in the rage of 1.5-2.0.

2 Description of the area covered by the study

An analysis of water needs was carried out for the following three districts in the Wrocław agglomeration: Oborniki Śląskie (the southern part of the District) – data for 2008, Wisznia Mała – data for 2007 and 2010 and Czernica – data for 2015. The districts are directly adjacent to Wrocław (Fig. 3).

In the Oborniki Śląskie District, places located in its southern part were analysed. This area is supplied with water by the Lubnów water supply system which covers the places: Lubnów, Niziny, Uraz, Raków, Kotowice, Paniowice, Zajączków and Pęgów, and by the Golędzinów water supply system supplying water to the places: Golędzinów and Nowosielce. In these water supply systems underground water is drawn from drilled wells and then it is treated and stored in clearwells. From the clearwells the water is pumped into the water-pipe network by water pumping stations. The network is built of PVC pipes 90-225 mm in diameter. In 2008 the considered area was inhabited by 4741 inhabitants. The water consumers are mainly single-family housing and services and to a small degree agriculture and industry.

Fig. 3. Area covered by the study.

The analysis of water consumption in the Wisznia Mała District was carried out in two stages. In the first stage the structure of the water needs in the places: Psary, Szymanów and Krzyżanowice (the Psary water supply system) and Kryniczno and Rogoż (the Kryniczno water supply system) was studied. In both the water supply systems underground water is drawn, which is subsequently treated and stored in clearwells and then pumped into the water-pipe networks by pumping stations. The network is built of PVC pipes 90-225 mm in diameter. In 2007 the population in the considered area amounted to 2,732 inhabitants. In the second stage investigations were carried out in the northern part of the District in the places: Machnice, Wysoki Kościół, Piotrkowiczki and Mienice (a pumping main) and Wisznia Mała, Ligota Piękna Pierwoszów, Malin, Strzeszów, Ozorowice and Szewce (a gravity main). The water-pipe system is supplied with water from the underground water intake in Pierwoszów and Machnice. The water-pipe network is built of cast-iron and PVC pipes 90-300 mm in diameter. In 2010 the places mentioned above were inhabited by 5,578 persons. The major water consumers in the whole analysed area are single-family housing and services. Water used for agricultural and industrial purposes amounted to a small percentage.

In 2015 studies were carried out in the Czernica District. The structure of water needs was determined for all the places located in the District, i.e. Nadolice Wielkie, Chrząstawa Mała, Chrząstawa Wielka, Nadolice Małe, Krzyków, Dobrzykowice, Kamieniec Wrocławski, Łany, Gajków, Jeszkowice, Czernica, Wojnowice and Ratowice. The particular places are supplied with water from the group water main with a water treatment plant in Nadolice Wielkie and from the Wrocław main. The water-pipe network is built of PVC and PE pipes 90-225 mm in diameter. The major water consumers are single-family housing and services, and industrial plants. Agriculture is a minor water consumer. In 2015 the population of the District amounted to 12,824 inhabitants.

In the studied areas the water supply network covers all the water consumers.

3 Research results

The results of the research into the water needs in the Oborniki Śląskie District, the Wisznia Mała District and the Czernica District are presented in Table 1 and illustrated in Fig. 4. The analyses were based on the data collected in the district organizational units administering the water supply systems described above.

Table 1. Summary of research results.

Place name	Population	Water demand, m^3/d		Average water consumption per inhabitant, q_j lpcd	Daily water demand irregularity coefficient, N_d
		$Q_{av.}$	Q_{maxd}		
Oborniki Śląskie District					
Golędzinów	462	39.65	89.98	85.8	2.27
Kotowice	380	43.11	96.19	113.4	2.23
Lubnów	520	40.00	100.63	76.9	2.52
Niziny	38	2.94	6.49	77.4	2.21
Nowosielce	51	3.33	7.92	65.3	2.38
Paniowice	307	28.43	71.27	92.6	2.51
Pęgów	1680	163.70	335.06	97.4	2.05
Raków	76	6.29	15.83	82.8	2.52
Uraz	900	59.87	132.91	66.5	2.22
Zajączków	327	23.05	55.21	70.5	2.40
TOTAL	4,741	410.37	911.49	86.6	2.22
Wisznia Mała District					
Psary	1006	107.67	207.81	107.0	1.93
Szymanów	728	80.27	159.39	110.3	1.99
Krzyżanowice	405	41.37	82.72	102.1	2.00
Kryniczno	362	58.36	155.97	161.2	2.67
Rogoż	231	22.47	43.58	97.3	1.94
TOTAL	2732	310.14	649.47	113.5	2.09
Ligota Piękna	904	104.83	282.62	116.0	2.70
Machnice	213	19.29	56.42	90.6	2.92
Malin	512	58.47	172.12	114.2	2.94
Mienice	265	28.12	85.20	106.1	3.03
Ozorowice	380	35.01	92.33	92.1	2.64
Pierwoszów	188	20.51	55.97	109.1	2.73
Piotrkowiczki	349	44.83	139.81	128.5	3.12
Strzeszów	1040	149.19	341.78	143.5	2.29
Szewce	875	100.93	251.16	115.3	2.49
Wisznia Mała	708	87.33	230.02	123.3	2.63
Wysoki Kościół	144	14.12	44.31	98.1	3.14
TOTAL	5,578	662.63	1,751.74	118.8	2.64
Czernica District					
Nadolice W.	1204	201.81	428.71	167.6	2.12
Chrząstawa M.	1028	121.17	247.39	117.9	2.04
Chrząstawa W.	1063	135.82	280.57	127.8	2.07
Nadolice Małe	246	36.11	76.67	146.8	2.12
Krzyków	371	56.59	121.32	152.5	2.14
Dobrzykowice	1477	289.71	633.14	196.1	2.19
Kamieniec Wr.	2336	330.21	680.36	141.4	2.06
Łany	181	27.41	49.80	151.4	1.82
Gajków	1091	158.15	318.24	145.0	2.01
Jeszkowice	910	108.93	219.03	119.7	2.01
Czernica	1257	147.18	279.50	117.1	1.90
Wojnowice	692	87.93	176.55	127.1	2.01
Ratowice	968	85.86	169.41	88.7	1.97
TOTAL	12,824	1,786.88	3,680.69	139.3	2.06

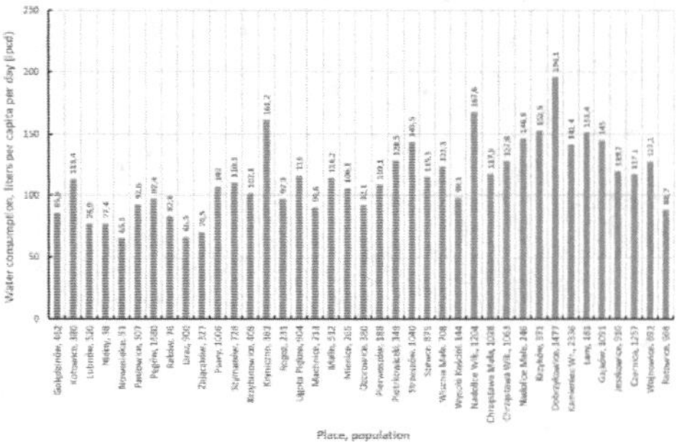

Fig. 4. Average unit water consumption in analysed places.

In the case of the parts of the Oborniki Śląskie District and the Wisznia Mała District, the collected data concerned the amount of water drawn for housing (households and watering home gardens), services, farms and livestock facilities, servicing motor vehicles and production plants. While in the Czernica District the average daily water consumption measured by water meters located at water service points was analysed with a special regard to the use of the drawn water.

The analyses show a wide variation of both the average unit water consumption, ranging from 65.3 to 196.1 lpcd, and the daily water demand irregularity coefficient, ranging from 1.82 to 3.14.

As regards average unit water consumption, the results for the places in the Oborniki Śląskie District, directly adjacent to Wrocław, coincide with the values given in the Minister of Infrastructure Order of 14 January 2002, whereas the ones for the places in the Wisznia Mała District and the Czernica District, especially in Kryniczno (q_j = 161.2 lpcd, N_d = 2.67), Nadolice Wielkie (q_j = 167.6 lpcd, N_d = 2.12) and Dobrzykowice (q_j = 196.1 lpcd, N_d = 2.19), are higher than the values published in the Order and in the trade journals. Various factors contribute to the relatively large water intake in the considered places, particularly in the Czernica District. First of all one should note the change in the use of mains water from rural to the one typical of urban housing estates, at a considerable use of water for purposes other than domestic, including watering lawns and greenery and the operation of swimming pools. Also quite significant is water abstraction for construction purposes.

4 Conclusions

The structure of the water needs in the water supply systems of the Oborniki Śląskie District, the Wisznia Mała District and the Czernica District, directly adjacent to Wrocław has been studied over the last decade.

The analyses have shown that various factors have a bearing on water intake variability. The determined average water consumption per inhabitant ranges from 65.3 to 196.1 lpcd while the daily water consumption irregularity coefficient ranges from 1.82 to 3.14. For most of the considered places, especially for the Czernica District, these values differ from the ones reported in the literature. It should be noted that this is due to the change in the character of the use of water by its consumers in the studied area from rural to the one typical of urban housing estates, with a considerable use of water for other than purely household purposes, such as watering lawns and greenery and the operation of swimming pools.

The observations indicate that it is necessary to continue the research on the variability of water consumption, particularly daily water consumption irregularity, in rural settlements adjacent to large cities in Poland.

This research was carried out as part of the statutory activity of the Faculty of Environmental Engineering at Wrocław University of Science and Technology, funded by the Ministry of Science and Higher Education in 2017 and 2018.

References

1. I. Zimoch, E. Łobos, Environ. Prot. Eng. **36**, 4 (2010)
   http://epe.pwr.wroc.pl/2010/zimoch_4-2010.pdf
2. Z. Siwoń, W. Cieżak, Instal Spec. edit. (2008)
3. A. Kotowski, *Foundations of the safe dimensioning of land drainages. Sewerage networks (Vol. I)* (Seidel-Przywecki Publishing House, Warsaw, 2015)
4. Z. Siwoń, J. Łomotowski, W. Cieżak, P. Licznar, J. Cieżak, *The analysis and prediction of water distribution in water supply systems* (in Polish). Polish Academy of Sciences. Civil Engineering Committee. Institute of Fundamental Technological Research, Warsaw (2008)
5. I. Zimoch, E. Łobos, Environ. Prot. Eng. **38**, 3 (2012)
   http://epe.pwr.wroc.pl/2012/3-2012/Zimoch_3 2012.pdf
6. J.L. Wescoat Jr, S. Fletcher, M. Novellino, Water Policy **18**, 4 (2016)
7. Municipal Infrastructure in 2014 and 2015 (in Polish). Central Statistical Office (2015, 2016)
8. E. Ogiołda, M. Kozaczek, Sci. Papers of Zielona Góra University **152**, (2013)
9. J. Pawełek, T. Bergel, O. Wojciechowska, Acta Sci. Pol. Formatio Circum. **14**, 4 (2015)
10. M. Orzechowska, JEENG **26**, (2011)
11. B. Tchórzewska-Cieślak, *12th All-Poland Scientific-Technical Conference, Issues of water supply and water disposal in agricultural-industrial regions* **30**, (2005)
12. J. Pawełek, G. Kaczor, Infrastrukt. Ekol. Ter. Wiejskich **2**, (2006)
13. H. Hotloś, Ochr. Srod. **32**, 3 (2010)

One or more water demand patterns? - Calibration of the selected water network model

Beata Kowalska[1,*], *Dariusz* Kowalski[1], *Paweł* Suchorab[1], and *Małgorzata* Iwanek[1]

[1]Lublin University of Technology, Faculty of Environmental Engineering, Nadbystrzycka 40B, 20-618 Lublin

Abstract. Calibration is one of the key steps in creating a numerical model of water supply networks. Calibration is performed on the basis of the results of the field measurement campaign and it should result in a consistent match of the simulation results with the results of the measurements. In the article the process and the results of the calibration of the selected water network model with the use a single (generalized) water demand pattern and using 11 patterns prepared for selected, characteristic consumers have been presented. Calibration was performed using the trial and error method in EPANET 2.0 and the optimization method in WaterGEMS.

1 Introduction

Increased complexity of water supply systems contributed to popularizing the use of professional computer software for managing a water supply company. Computer models support planning, designing, modernizing and operating vast water supply networks [1-3]. The quality of simulation analyses largely depends on the quality of model design, which is reflected in: representation of the structure of particular elements, identification of water supply system elements, water demand models and the parameters of water supply network [4,5]. The required degree of model simplification depends on the goal of simulations (hydraulic conditions, quality changes, energy cost). In the case of the existing networks, a model must be calibrated to function properly. Calibration involves adjusting the parameters of particular model elements, so that the obtained calculation results correspond to the actual conditions in a network [6,7]. During the calibration of hydraulic models, the following parameters may be subjected to adjustment: pipe roughness coefficient, values and patterns of water demand and status of gate valves. Thus far, there are no unequivocal matching criteria for the above-mentioned parameters. Certain recommendations pertaining to static and dynamic models were devised in the United Kingdom and the United States [8]. The difficulties in achieving a match between the results obtained on the basis of a model and the results from measurements may stem from the uncertainty of model structure, uncertainty of the calculation method as well as the uncertainty of the devices used for measurement and control.

One of the most important factors having an influence on the simulation calculations is the correct allocation of water demand in particular junctions of the hydraulic model. This is a difficult task due to the variability of water demand in various time cycles and the diversity of their distribution depending on the recipient category. Correct inclusion of water demand changes by various recipient categories is especially important in Extended Period Simulations (EPS) models.

The temporal variability of water demand may be analyzed over multiple years, annually, seasonally, monthly, weekly, as well as daily and hourly. Diversification both in the daily and hourly water demand is dependent on numerous factors. They include the regional diversification, economic conditions, sociological factors as well as highly important demographic status and the industrial development in a given region [9].

The presented work made an attempt of calibrating an existing hydraulic model of a selected water supply network with the use of individual water demand patterns devised for the selected recipients and comparison of the results with the ones obtained from the calibration conducted with a general pattern. The calculations were carried out in EPANET 2.0 (US EPA) and WaterGEMS software by Bentley. The assessment of matching was additionally conducted with the use of statistical analyses performed with Statistica 13.1 software package by StatSoftPolska.

2 Object description and research methodology

The analyzed water supply network supplies approximately 53 thousand residents and about 4300 economic entities, mainly representing services, trade, gastronomy, administration centers, education and health care. The water supply network uses groundwater, the well water is supplied untreated to two field storage tanks. Then, it is transmitted via suction pipeline with the diameter of Ø300 and supplied to the water supply

*Corresponding author: b.kowalska@pollub.pl

network by means of a pumping station. The water supply pipelines, with the length of roughly 170 km are mainly made of grey cast iron and ductile iron, asbestos cement, steel, polyethylene and polyvinyl chloride.

The conducted analyses were based on the numerical model of the water supply network prepared earlier, presenting the geometrical structure of this network, location of objects, diameters, pipe roughness, as well as levels of foundation. The model was devised as "basic", i.e. containing all the water mains and branches, without the house connections. It comprises 2785 junctions and 2987 connecting pipes. The included water demand corresponds to an average daily value. The daily distribution of water demand (general pattern) was devised on the basis of readouts from two flow meters located at the pumping station outlet.

In order to devise the data pertaining to individual water recipients, following the analysis of water recipient register and discussions with local Municipal Water and Sewerage Company, 11 categories were assumed: shop, kindergarten, sports hall, high multi-family buildings, medium multi-family buildings, shopping centre, office, hotel, detached houses, terrace houses, small production plant with low water demand. The patterns were devised as part of diploma papers [10,11]. Owing to the applied electronic devices for remote radio readout of water meters, it was possible to collect data for 48 hours with hourly intervals. The study lasted for 15 days, including two weekends.

For the purpose of the hydraulic model calibration, pressure measurements were carried out in 15 measurement locations. The pressure meters were installed on selected fire hydrants and one was placed at the pumping station outlet. The pressure in the selected measurement locations was taken using Cell-Box-H meters, equipped with loggers saving the data. The flow rate in 9 locations was measured using Porta Flow 300, Rota Yokugawa 80 and HydrINSultrasound flowmeters. Additionally, the existing water- and flowmeters, installed permanently in the pumping station, were used. At the first stage of analyses, the hydraulic model built using EPANET 2.0 software was calibrated with the general pattern; then, individual patterns prepared for the afore-mentioned 11 recipient categories were introduced and the model was recalibrated in EPANET 2.0. Afterwards, the hydraulic model with introduced individual patterns was converted to WaterGEMS software and again calibrated with the available tools.

During the calibration of the hydraulic model of a water supply network in EPANET 2.0, an integrated calibration report module was utilized [12]. The calibration process was carried out using a trial and error method, which involved an attempt to obtain the best match between the results of measurement and model calculation. In this method, the pipe roughness, pressure at the pumping station outlet and the parameters of pressure-reducing gate valves were adjusted "manually". The calculated water demand was adjusted to the supply measured in the pumping station through a uniform daily peaking factor.

Calibration of the hydraulic model in WaterGEMS software by Bentley, was conducted in the proprietary Darwin Calibrator module [13]. In this case, the automatic calibration utilizing the genetic algorithm with three calibration criteria, including: Minimize Difference Squares; Minimize Difference Absolute Values and Minimize Maximum Difference, was selected. Calibration was carried out based on the same pressure and flow rate measurement results used in EPANET 2.0. The calibration involved adjusting the pipe roughness coefficient (ROUGHNESS) and the water demand in a junction (DEMAND). In the genetic algorithm, the variability of pipe roughness and junction water demand ranging from 0.5 to 1.5 of the initial value, was assumed.

3 Research results and discussion

The first stage of research involved calibration of the hydraulic model (pressure and flow rate) in EPANET 2.0 software. The calibration was conducted using a single, general pattern of water demand; then, individual patterns (prepared on the basis of 11 characteristic types of recipients) were introduced and the calibration was repeated. When the calculations were finished, a combined calibration report was generated for the analyzed cases. The obtained coefficients of determination r^2, reflecting the matching degree of linear regression to the data, virtually did not differ from one another. Their values, exceeding 0.9 suggest a very good match. However, one should remember that the quality evaluation of the calibration obtained with EPANET 2.0 software does not reflect the actual matching degree of model parameters, because it involves the mean values.

The next step included calibration in WaterGEMS software, which utilized converted hydraulic model with introduced individual patterns and a genetic algorithm. The coefficient of determination obtained on the basis of non-averaged flow rate measurement spots, reached $r^2 = 0.872$ and reflected the actual conditions in the analyzed water supply network better. For the sake of comparison, the r^2 for mean values was calculated as well, yielding a nearly identical value to EPANET 2.0 (Table 1). This proves that averaging does not reflect the actual matching degree of the model to the measured values.

Table 1. Comparison of r^2 after calibration.

Parameter	EPANET 2.0		WaterGEMS	
	General pattern	Individual patterns		
	r^2 for			
	mean values of parameters			not averaged parameters
Flow rate	0.956	0.967	0.966	0.872
Pressure	0.998	0.998	0.999	0.980

The coefficient of determination obtained on the basis of all measurements (without averaging) – enables a general assessment of the quality of conducted calibration. In order to perform more detailed analyses, calculations were carried out in the measurement locations in which pressure and flow rate were measured during a campaign. The differences between the calculated values and the ones measured in particular locations could be observed on the graphs of these changes in time. Fig. 1 and Fig. 2 present the results for exemplary locations (junction, pipe) designated in the model as J2502 and P2417. Both the junction and the pipe represent "average" differences between the calculated and measured values. In the remaining locations the differences were lower, sometimes almost indiscernible, or higher from the presented values.

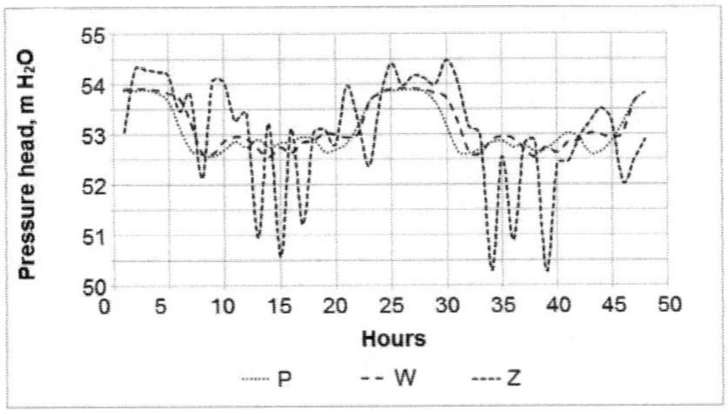

Fig. 1. Comparison of the results of simulation of pressure head changes in node J2502 using general pattern (P) and individual patterns (W) with measurement results (Z).

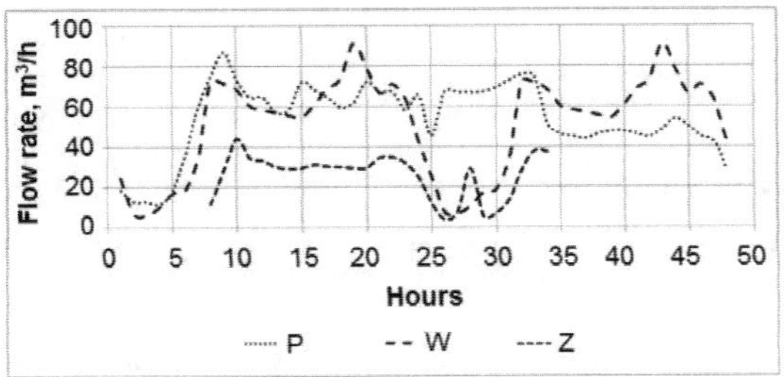

Fig. 2. Comparison of results of flow rate simulations in pipe P2417 using general pattern (P) and individual patterns (W) with measurement results (Z).

In order to picture the matching degree between the measured and calculated values in the measurement locations for different types of patterns (general and individual), statistical analyses were performed in Statistica 13.1 software package. The Figs 3, 4 and 5 present the matching degree of the pressure in junction J2502 analyzed earlier, following calibration in EPANET 2.0 and WaterGEMS.

The value of the coefficient of correlation while using the general pattern was $r = 0.38654$ and was lower than the value obtained after the introduction of individual patterns ($r = 0.52647$ EPANET 2.0, $r = 0.63866$ WaterGEMS). Therefore, making the water demand patterns more specific contributed to positive results.

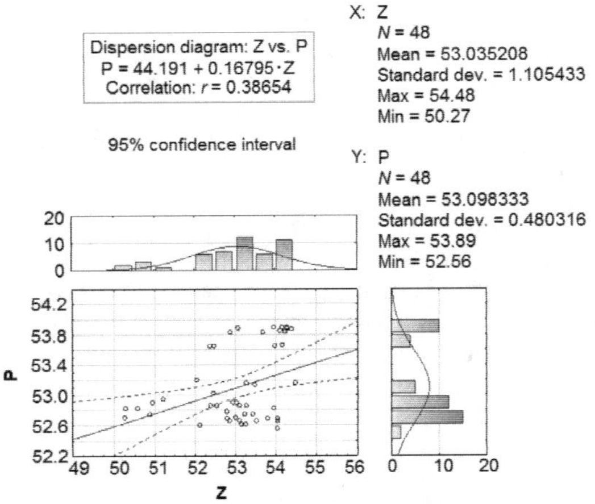

Fig. 3. Comparison of the results of the simulations of pressure head changes in node J2502 using general pattern (P) with measurement results (Z).

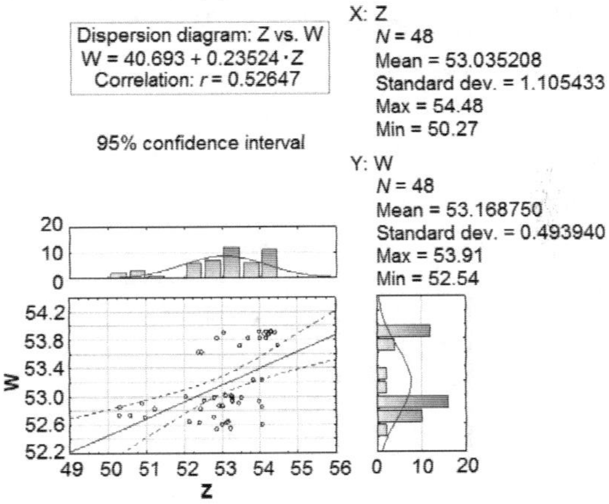

Fig. 4. Comparison of results of simulation of pressure head changes in node J2502 using individual patterns (W) with measurement results (Z) (EPANET 2.0).

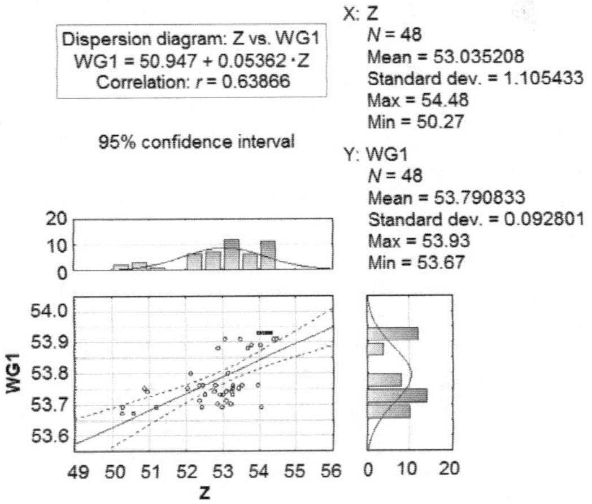

Fig. 5. Comparison of results of simulation of pressure head changes in node J2502 using individual patterns (WG1) with measurement results (Z) (WaterGEMS).

However, taking into account all measurement locations, it is difficult to unequivocally evaluate the influence of implemented changes, because in some cases the results were almost the same.
The Fig. 6 presents (as an example) the matching degree of flow rate in pipe P2417 (with individual patterns), which was also analyzed earlier. Using the general pattern, virtually no correlation was observed. Similarly to the previous case, it is difficult to unequivocally evaluate the influence of implementing individual patterns, because in other junctions, the situation varied; sometimes the differences were almost indistinguishable.

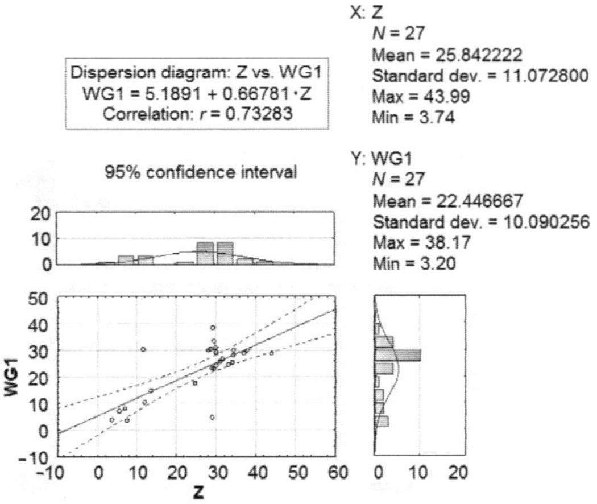

Fig. 6. Comparison of flow rate simulation results in pipe P2417 using individual patterns (WG1) with measurement results (Z) (WaterGEMS).

4 Conclusions

• Taking into account the results of analyses conducted in the work, which aimed at evaluating the impact of applying individual water demand patterns on the calibration results, it can be stated that adding more details to the data introduced to the hydraulic model seems to be a right direction.
• Unfortunately, the calibration module of EPANET 2.0 software does not enable a correct evaluation of the matching degree of the analyzed variables. The reported values of coefficient of determination r^2 are unreliable, because they pertain to the averaged values; therefore, it was virtually impossible to observe the difference after the implementation of changes in the model under the conditions of considered water supply network.
• Application of the automatic calibration, utilizing genetic algorithms for calculations, which is included in WaterGEMS software, enabled to reflect the actual conditions in a water supply networkbetter. Moreover, the possibility of using the automatic calibration module facilitates and accelerates the calibration process, which is required each time changes are introduced to the existing model.
• The statistical analysesconducted using Statistica 13.1 software package in particular junctions of the model enabled to evaluate the differences in matching of the model data following the implementation of more detailed, individual water demand patterns. Slightly better results were obtained after the calibration in WaterGEMS. The fact that the obtained results were not always unequivocal could have stemmed, i.a. from the inconsistencies and omissions during the measurement campaign, e.g. lack of remote water meter readouts, which prevented a periodic synchronization of the water demand values with the values measured in the network. Hydrant tests were not carried out during the pressure measurementseither.
• To sum up, the obtained results confirm that calibration is an extremely difficult and complex process, and its quality depends on numerous factors. The obtained findings enabled to observe a slight improvement in the results following the introduction of more detailed individual water demand patterns, and thus may constitute the basis for further research, which should be verified in terms of the observed issues.

The Polish extended version of the paper was published in the journal *Instal* 7–8, 2017, p. 57–60.

References

1. I. Zimoch, Ochr. Środ. **30**, 3 (2008) http://www.os.not.pl/docs/czasopismo/2008/Zimoch_3-2008.pdf
2. A. Bałut, J. Bylka, Instal **12**, (2013)
3. I. Zimoch, Ochr. Środ. **34**, 4 (2012)
4. A. Kotowski, A. Pawlak, P. Wójtowicz, Ochr. Środ. **32**, 2 (2010)
5. I. Zimoch, E. Łobos, Environ. Prot. Eng. **36**, 4 (2010) http://epe.pwr.wroc.pl/2010/zimoch_4-2010.pdf
6. Z. Siwoń, W. Cieżak, Instal **1,** (2008)
7. Z. Kapelan, *Calibration of Water Distribution System Hydraulic Models* (PhD thesis, University of Exeter 2002)
8. AWWA Engineering Computer Applications Committee, *Calibration Guidelines for water distribution system modeling* (1999), Reprinted from *Proceedings of the AWWA Information Management and Technology Conference, New Orlean, Louisiana, April 1999*
9. E. Kozłowski, B. Kowalska, D. Kowalski, D. Mazurkiewicz, Arch. Civ. Mech. Eng. **18,** 1 (2018)
10. J. Połeć, *Ocena warunków hydraulicznych wybranej strefy wodociągu z uwzględnieniem odbiorców indywidualnych* (MSc thesis, Lublin University of Technology, 2016)
11. P. Wyglądała-Szarek, *Analiza warunków hydraulicznych w oparciu o wzorce indywidualne we fragmencie miejskiej sieci wodociągowej* (MSc thesis, Lublin University of Technology, 2016)
12. A.L. Rossman, *EPANET 2 Users Manual* (U.S. EPA, Cincinnati 2000)
13. T. Walski, D. Chase, D. Savič, W. Grayman, S. Beckwith, E. Koelle, *Advanced Water Distribution Modeling and Management* (Bentley Institute Press, Exton, Pennsylvania USA, 2007)

Application of K-nearest neighbours method for water pipes failure frequency assessment

Małgorzata Kutyłowska[1,*]

[1]Wrocław University of Science and Technology, Faculty of Environmental Engineering, Wyb. Wyspianskiego 27, 50-370 Wroclaw, Poland

Abstract. The paper describes the results of failure rate modeling using K-nearest neighbours method (KNN). This algorithm is one among other regression methods, called machine learning methods. The aim of the presented paper was to check the possibilities of application of such kind of modelling and the comparison between current results and investigations of failure rate prediction in another Polish city. Operational data from 12 years of exploitation, received from water utility, were used to predict dependent variable (failure rate). Data (249 and 294 for distribution pipes and house connections, respectively) from the time span 2001–2012 were used for creating the KNN models. On the basis of other data (one case for each year) the validation of optimal model, based on Euclidean distance metric with the number of nearest neighbours K = 2, was carried out. The realization of the modelling was performed in the software program Statistica 12.0.

1 Introduction

The assessment and analysis of failure frequency and reliability of water distribution systems and their elements have been very precisely described in the literature [1–3]. Nowadays, typical approach concerning the analysis of the number and kinds of occurring damages should be and is extended of the effects of water losses [4] which should be minimalized due to limited water resources in Poland. Moreover, more often the effects of failures are evaluated by consumers [5] and according to their opinion it is required to improve the quality of water delivering.

1.1 Modeling

The investigations related to the technical condition of buried infrastructure are extended and many mathematical modeling methods and special software are used to increase the efficiency of the management of water-pipe networks [6–8]. The prediction of failure rate could be carried out not only using typical statistical methods [9], but also using algorithms based on artificial intelligence [10–12]. Artificial intelligence known as neural networks based on unsupervised learning (Kohonen network) could be successfully used for risk analysis of water distribution system [13]. Till now K-nearest neighbours (KNN) method was not widely used for reliability assessment of municipal systems and prediction of selected indicators. The main aim of this work was to indicate the possibilities of using KNN algorithm for failure analysis of water conduits. This method was successfully applied in many scientific fields, e.g. for failure analysis of mechanical facilities [14] and in broadly understood medicine [15]. Hence it seems to be reasonable to check the usability of KNN method in failure frequency analysis of water supply systems.

1.2 K-nearest neighbours method

K-nearest neighbours algorithm is relatively easy in implementation and in analysis in comparison to other regression methods. KNN could be used in classification [16] or regression [17] problems. Modeling of failure rate of water pipes is based on regression not classification algorithm. The main assumption of this algorithm is to classify similar data to the same classes. The prediction of dependent variable is based on the comparison if this variable belongs to the exemplary set or not [18]. The choice of the number of K-nearest neighbours has the great influence on the model quality. The lower number of *K*, the bigger prediction errors. On the other hand, too many K-nearest neighbours can lead to modeling results with so-called error of overfitting. Similarly, as in other regression methodologies, it is required to find the optimal solution. Optimal number of *K* is not known *a priori*. The application of so-called V-fold cross validation is recommended for finding the most beneficial result in terms of the model quality and the agreement between dependent variable and experimental values. In this type of cross validation, data are divided into V randomly selected disjoint parts. Using the V-1 parts of data as training examples the dependent variable is predicted and the prediction error is calculated on the basis the residual sum of

* Corresponding author: malgorzata.kutylowska@pwr.edu.pl

https://doi.org/10.1051/e3sconf/20185900021

squares. The procedure is executed for all the V data segments. Then a model quality measure is determined on the basis of the averaged errors of the particular cycles. The optimal model parameters are selected during a quality analysis [18]. In regression problems, the average for K nearest neighbours is calculated according to the equation (1) [18]:

$$y = \frac{1}{K}\sum_{i=1}^{N} y_i \tag{1}$$

where y_i is the output value for i learning example and y is the value of output variable for new example. The result is obtained on the base of the K nearest neighbours of new point. Following this assumption, it is needed to have some kind of measurement of the distance between examples. There are four types of distance metric: Euclidean (D_E) – equation (2), quadratic Euclidean (D_{E2}) – equation (3), Manhattan (D_M) – equation (4) and Czebyszew (D_C) – Equation (5) [18]:

$$D_E(x_n, p) = \sum_{i=1}^{N} \sqrt{(x_{ni} - p_i)^2} \tag{2}$$

$$D_{E2}(x_n, p) = \sum_{i=1}^{N} (x_{ni} - p_i)^2 \tag{3}$$

$$D_M(x_n, p) = |x_{ni} - p_i| \tag{4}$$

$$D_C(x_n, p) = \max_{1 \le i \le N}(|x_{ni} - p_i|) \tag{5}$$

where $D(x_n, p)$ is the distance metric, x_n is the new point and p is the learning example. The regression or classification precision depends mainly on the metric used to calculate distances [19].

2 Methodology of studies

The calculations were carried out in the programme Statistica 12.0. Operating data, received from water utility, from the time span 2001–2012 were used for modeling separately (with different models) failure rates (λ, fail./(km·a)) of distribution pipes (DP) and house connections (HC). The whole data set (249 data for distribution pipes and 294 data for house connections) was randomly divided into two samples (learning – 75% and testing – 25%). Moreover, the model validation (verification) was carried out using data which did not belong to mentioned above two sets (one case for each year). KNN models were created on the basis of all four kinds of distance metrics (D_E, D_{E2}, D_M and D_C). Indicators (dependent variables) λ_r for distribution pipes and λ_p for house connections were predicted using independent variables. Independent variable (so-called predictors) vector contained (separately for distribution pipes and house connections): length-L, diameter-D and material-M (cast iron-Z, steel-S, galvanized steel-SO, PE, PVC) as well as the year of construction-YC of water pipes. The short description of investigated water-pipe network was presented in the work [20]. In the current analysis 10-fold cross validation method was used. Such approach was also applied in other regression algorithms, e.g. in regression trees. The experience of the precursor of regression tree methodology, Breiman et al. [21] indicated that just V = 10 is the optimal value. The range of experimental dependent and independent variables in years 2001–2012 is displayed in Table 1 (learning and testing sample). The number of failures of house connections as well as of distribution pipes in the learning sample varied respectively between 10 and 27 as well as 8 and 32 in the time span of 2001-2012.

Table 1. Dependent and independent variables – learning and testing.

	L, km	D, mm	YC	λ, fail./(km·a)
DP	57.3–88.7	80–200	1961–2006	0.10–0.57
HC	23.4–50.2	20–100	1961–2012	0.23–1.59

3 Results and discussion

In Fig. 1 and Fig. 2 the real (experimental) and predicted (by KNN model) failure rates of house connections and distribution pipes are displayed. The results of modeling concern the model validation step. Such approach seems to be reasonable, because the model quality assessment should be carried out using the data which were not included to the learning and testing step of modeling. For learning and testing step the prediction results were very good. Pearson correlation coefficient was established at the level of ca. 0.99.

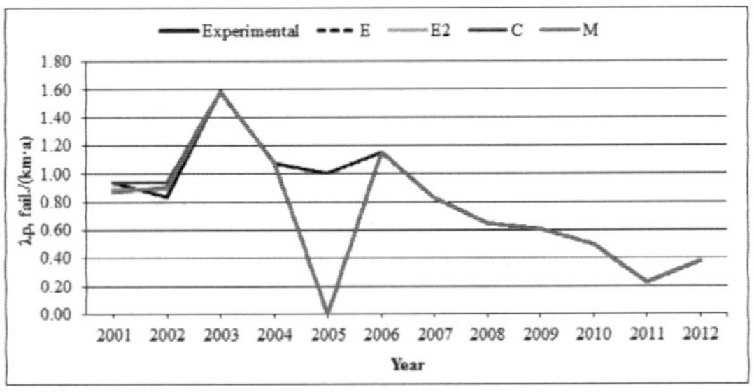

Fig. 1. Experimental and predicted failure rate of house connections.

The results of prediction of indicator λ for house connections (Fig. 1) are almost ideal convergent with real values for all kinds of distance metric. Some small differences between experimental and predicted failure rates (for all distance metrics) are visible for years 2001 and 2002. On the other hand, completely not understood is the result for 2005 when the model generated value of indicator λ equalled to zero for all distance metrics (D_E, D_{E2}, D_M and D_C). Operational data for year 2005 could not be treated as outliers in comparison to other analysed years. It could be only assumption that modeling using K-nearest neighbours method contains the elements of so-called „black box" (similarly as other machine learning algorithms [22], which are also classified to regression methods). It means that deep inquire in the way of creating the optimal model and optimal results is rather impossible.

More surprising results of validation of the KNN model are observed for prediction of failure rate of distribution pipes (Fig. 2). In this case the ideal convergence between predicted and experimental values for each analysed year and all kinds of distance metric was achieved (in the Fig. 2 series KNN means that all distance metrics gave the same results).

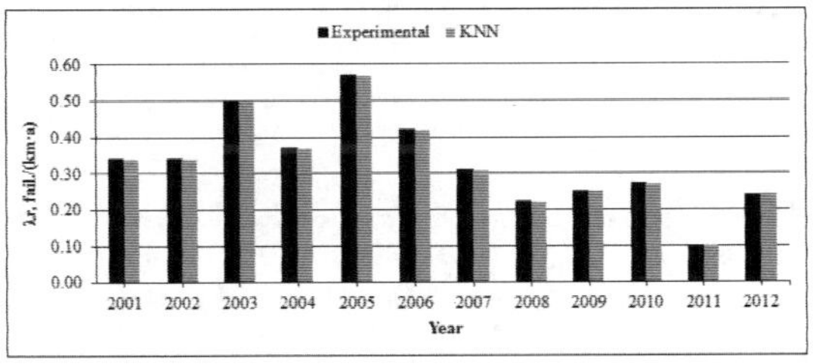

Fig. 2. Experimental and predicted failure rate of distribution pipes.

Completely different results were observed during the failure analysis also using KNN model on the basis of operational data from another Polish city [23]. It is obvious that each kind of modelling and forecasting is and even should be burden by the prognosis error which results from not only the data inaccuracy, but also from the difficulties in describing the relations between dependent (λ) and predictors. Hence it is required to be careful when such ideal results as above are observed. Ideal agreement between modeling results and real values should raise doubts due to many limitations of mathematical modeling connected with the character of changing dynamically the natural phenomena describing the technical condition and failure frequency of water distribution system. Maybe the differences in modeling results displayed in this work in comparison to solutions obtained earlier [23] are connected with the kinds of predictors as well as with the size of the independent variable vector. The analyses of the results in both investigations indicate that the kind and number of independent variables have the great influence on the modeling quality. This aspect is maybe even more important than the number of neighbours *K* or the kind of distance metric. Moreover, there is one more difference in the approach presented in this work in comparison to previous modeling attempts by means of KNN [23]. The kind of the created model is significant issue. In this work two separate models for failure rate prediction of distribution pipes and house connections were built. In contrast, in another author investigations [23] one model was analysed. This one model had three dependent variables: indicators not only λ_p, λ_r (for distribution pipes and house connections) but also indicator λ_m (water mains). Despite the fact that in currently analysed case different independent variables, which in better way describe the water conduits, were used, created models seem to be simpler than model with three output variables. This fact is obviously connected with the modeling quality. More complicated models (with more complicated structure) could be burden by higher prediction errors than models with different predictors but with one dependent variable in the output.

The changes of cross validation errors values in the contrast with the number of nearest neighbours are displayed in the Fig. 3. The optimal number of *K* is also indicated.

a)

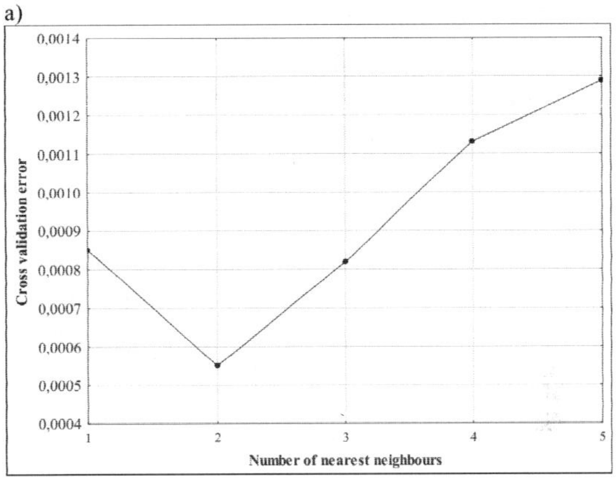

b)

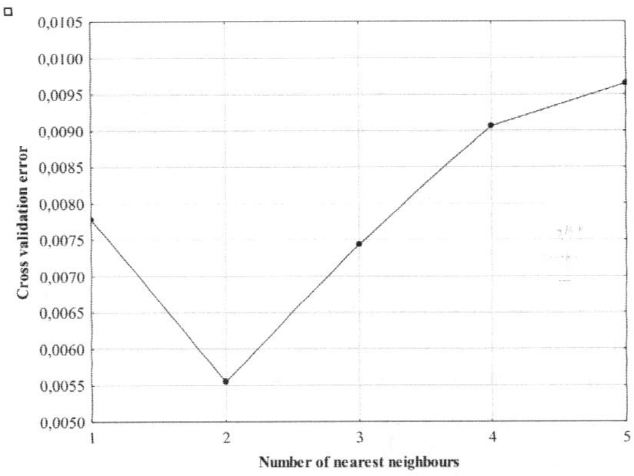

Fig. 3. Changes of cross validation error vs. number of K, a) distribution pipes, b) house connections.

Concerning distribution pipes and house connections, the optimal number of nearest neighbours is the same and equals $K = 2$. The maximum number of nearest neighbours (5), determined during the model creation, depends on the number of independent variables and on the number of learning data. The analysis of the figure 3 indicates that the lowest value of the cross validation terror (0.00555 – house connections and 0.00055 – distribution pipes) was observed in the models described by Euclidean (D_E) distance metric. Moreover, when the minimal values were achieved, the error value was increasing with the increase of the number of K. Despite the fact that the prediction results (Fig. 1 and 2) do not differ in relation to distance metric, it seems that Euclidean distance metric is the optimal one for describing the failure rates of house connection as well as for distribution pipes. Application of other distance metric had the great influence on the higher number of K, what meant that the complexity of the model was bigger. The architecture (structure) of the model should be as simple as possible. In addition in many other modeling, by means of KNN method, problems just Euclidean metric was proposed as the optimal one [19]. This distance metric describes relatively good the relations between dependent variables and predictors, especially when this dependence is not known *a priori* [19].

4 Conclusions

The main aim of this work was to verify if application of K-nearest neighbours method is possible for prediction of failure rates of water pipes. The modeling, on the basis of operational data from the time span 2001–2012 received from one water utility, of the indicator λ (separately for distribution pipes and house connections) was performed using following independent variables: material, diameter, length and year of construction of water conduits. Models were created based on all kinds of distance metric (Euclidean, quadratic Euclidean Manhattan and Czebyszew). The maximum number of nearest neighbours was established at the level of $K = 5$. The increase of the number of K leads to increase of 10-fold cross validation error. The validation results of optimal models based on Euclidean distance metric with $K = 2$ are satisfactory. The ideal convergence between predicted and experimental values of failure rate was obtained. On the other hand, one should be very careful and sceptical when such great correlation is observed, because this fact could denote that model was overtraining. In such case the predicted values are ideally fit to learning sample what could mean that the generalization capabilities are lost. On the basis of the results shown in this paper one can conclude that KNN method seems to be less appropriable for failure rate prediction of water-pipe network than other regression algorithms (so-called learning machine methods), e.g. artificial neural networks, regression tress and support vector machine. The modeling does not mean the ideal agreement between the experimental and predicted values of variables, but rather means the recognition of the relationships between predictors and dependent variable. Many studies conducted until now in the whole world show that KNN method has not been applied for analysis and the assessment of failure frequency of water conduits. That is the reason to use this algorithm in the failure analysis of water-pipe network in Poland. It should be indicated that the choice of number and kinds of independent variables as well as the whole modeling process were carried out with the knowledge about KNN in relation to other engineering issues. In such case it is necessary to deepen the modeling methodology in relations exactly to investigations connected directly with reliability analysis of municipal systems. Maybe, this hypothesis should be checked using operational data from other water distribution systems, KNN method, in distinction to other regression algorithms, does not require too large (number of cases and variables) vector of predictors. Generalization ability maybe has been just lost due to using ca. 200 learning cases. If such hypothesis is confirmed during the modeling of failure rate of water pipes in another city, it will mean that KNN algorithm is privileged to other regression methods on account of difficulties of obtaining huge operating data.

The work was realized within the allocation No. 0401/0069/16 awarded for Faculty of Environmental Engineering Wroclaw University of Science and Technology by Ministry of Science and Higher Education in years 2016–2017.

References

1. H. Hotloś, *Quantitative assessment of the effect of some factors on the parameters and operating costs of water-pipe networks* (Wrocław University of Technology Publishing House, Wrocław, 2007)
2. M. Kwietniewski, J. Rak, *Reliability of water supply and wastewater infrastructure in Poland* (Polish Academy of Science, Warszawa, 2010)
3. I. Zimoch, Ochr. Sr. **34**, 4 (2012)
4. M. Iwanek, A. Musz, B. Kowalska, D. Kowalski, M. Chołody, Instal **1** (2016)
5. J. Rak, K. Pietrucha-Urbanik, Instal **2** (2016)
6. B. Tchórzewska-Cieślak, D. Szpak, Ochr. Sr. **37**, 3 (2015)
7. A. Scheidegger, J.P. Leitao, L. Scholten, Water Res. **83** (2015)
8. I. Zimoch, Ochr. Sr. **30**, 3 (2008)
9. Y. Kleiner, B. Rajani, Urban Water **3**, 3 (2001)
10. M. Kutyłowska, Instal **1** (2016)
11. M. Kutyłowska, Period. Polytech-Civ. **61**, 1 (2017)
12. M. Nishiyama, Y. Filion, Can. J. Civil Eng. **41**, 10 (2014)
13. K. Pietrucha-Urbanik, B. Tchórzewska-Cieślak, *Water Supply System operation regarding consumer safety using Kohonen neural network* (in Safety, Reliability And Risk Analysis: Beyond The Horizon, 2014)
14. A.B. Andre, E. Beltrame, J. Wainer, Appl. Artif. Intell. **27**, 1 (2013)
15. H. He, W. Graco, X. Yao, *Application of genetic algorithm and K-nearest neighbour method in medical fraud detection* (in SEAL'98, Springer-Verlag, 1999)
16. C.H. Wana, L.H. Lee, R. Rajkumar, D. Isa, Expert Syst. Appl. **39**, 15 (2012)
17. A. Mangaro, F. Pizzo, A. Lombardo, A. Pogliaghi, E. Benfenati, Chemosphere, **144** (2016)

18. Statistica 12.0, Electronic Manual
19. K.Q. Weinberger, L.K. Saul, J. Mach. Learn Res. **10** (2009)
20. M. Kutyłowska, M. Orłowska-Szostak, Water Practice and Tech. **11**, 1 (2016)
21. L. Breiman, J.H. Friedman, R.A. Olshen, C.J. Stone, *Classification and regression tress* (Chapman&Hall/CRC, Boca Raton, 1984)
22. S.H. Hamraz, S.S. Feyzabadi, *General-purpose learning machine using K-nearest neighbors algorithm* (in RoboCup 2005, Springer-Verlag, 2006)
23. M. Kutyłowska, *Prediction of failure rate of water pipes using K-nearest neighbours method* (in IWA 8th Eastern European Young Water Professionals Conference, 2016)

Methodical basis of the needs of water supply in rural areas in normal and special conditions

Mikołaj Sikorski[1], and *Hanna* Bauman-Kaszubska[1,*]

[1]Faculty of Civil Engineering, Mechanics and Petrochemistry, The Warsaw University of Technology, Łukasiewicza 17, 09-400 Płock, Poland

Abstract. When calculating the balance of water supply, the purpose for which water is intended should be taken into account. Depending on them, the water quality parameters may vary. Rural and agricultural water demand covers the basic types of water demand, including the population's living and economic needs, animal husbandry, the needs of public utilities, the needs related to the operation of vehicles and machinery, workshops, machines and other purposes, including the own needs of the water pipes, fire-fighting etc. The level of demand is also closely related to the factors influencing the level of individual water consumption. Taking into account the deficiencies in formal and legal regulations, the binding regulations concerning the operation of water supply systems in special conditions have been presented so far. Elements of the benchmarking study on unit water demand indicators in normal and special conditions in rural areas have also been taken into account, guided by the principles and numerical indicators for the calculation of water demand for drinking and business purposes.

1 Introduction

The role and importance of water is by its users usually adopted as the fulfillment of basic living and economic needs. This awareness, based on daily living, generally increases in cases of lack of access to water sources and the need to bring it sometimes from quite considerable distances, as well as in emergency situations, natural disasters and other hazards.

Waste and lack of respect for water are accompanied by easy and unrestricted access to it. Contaminated water used to water animals can cause many illnesses and even collapsing. Chemicals such as nitrates and pathogens (bacteria, viruses, worms, etc.) that cause poisoning and animal diseases are particularly dangerous.

It can be said that water is a basic and indispensable factor for the health and well-being of a farmer and his family and the basic factor of agricultural, vegetal, animal and processing production. Only clean water of good quality is useful.

The basic element in water management, including rural areas, is to save water consumption and prevent water pollution, regardless of where they occur.

The water supply of rural areas is characterized by many water requirements, including, among others, the needs of living and economic purposes, animal husbandry, public utilities, agricultural and agri-food processing plants, washing of agricultural equipment and vehicles, workshops as well as other needs.

For rural settlement units and agricultural centres, the demand for animal husbandry and living and economic purposes are the decisive factors that play a fundamental role in shaping water abstraction, and thus affecting the size of technological facilities in water treatment plants. The related unit water demand rates depend on many factors, including the standard of sanitation of households, the equipping of households with sanitary utensils, the equipping of these units with collective or individual sanitation facilities, as well as the size of animal husbandry, the size and intensity of animal production, and many other factors.

The functioning of the water supply system is connected with the possibility of threatening human health or life, and the failures that occur often lead to catastrophic consequences [1]. As a part of the water supply system security, it is necessary to monitor continuously operational statuses and identify adverse events together with procedures for responding to such events [2].

2 Legal and formal bases

Water management issues in the country are governed generally by the Water Law Act of 18 July 2001 [3], as well as on the technical and organizational side by the Act of 7 June 2001 on Collective Water Supply and Collective Sewage Disposal [4] and Legal acts related to these laws.

Basic legal regulations related to the above issues have undergone a kind of evolution, which was influenced, among other things, by the actual amounts of water consumption for the objectives, charges for water intake, as well as rational conditions justifying the need to unify the scale of unit water demand indicators at the

* Corresponding author: Hanna.Bauman@pw.edu.pl

national level. For the purposes of drawing up the balance of demand and water consumption under normal conditions, the Regulation of the Minister of Infrastructure of 14 January 2002 on average water consumption standards [5] shall apply. It should be noted that in the so-called special conditions, the amount of water demand and consumption is determined in accordance with the following legal regulations, i.e. in:

• Regulation of the Minister of Spatial Economy and Construction of 21 September 1995 on the principles of ensuring the functioning of public water supply facilities in special conditions [6],

• Regulation of the Council of Ministers of 25 June 2002 on the detailed scope of activities of the Chief of the National Civil Protection, chiefs of civil defense of voivodships, countries and communes [7].

This issue was also addressed in publications [8-12]. These are the experiences of national science and practice related to the role of water in human and animal life.

The aforementioned MSEC decree [6] and its amendment, as laid down in the 2002 Council of Ministers decree [7] concerns the following:

• municipal and rural waterworks,
• industrial waterworks,
• emergency wells.

A limited period of water supply is defined as the condition during which reduced water supplies occur, with a possible decrease in quality parameters. In the period of limited water supply caused by external factors (failures, natural disasters, state of danger and possible war, etc.) one should consider:

• the necessary delivery period of several weeks,
• the minimum delivery period of several days.

The necessary amount of water supplied during a limited supply period is determined by the amount of water that covers the needs of all consumers in the quantities necessary for life, functioning of public utility plants and the needs of the selected production. The minimum quantity of water supplied during a limited supply period, on the other hand, is defined as the quantity of water that covers the survival needs of the population and livestock. Public wells should be understood as wells made on the basis of the Ordinance of the Council of Ministers of 23 December 1986 on water supply facilities and sewerage facilities, as well as water charges and wastewater disposal charges [13].

3 Unit water demand ratios

Available bibliographic data [14-17] demonstrate the occurrence of much smaller unit water consumption from those given by the guidelines and the continued downward trend in both individual farms and entire water supply systems. There is therefore a need for research in rural areas, not only in typical agricultural holdings, but also in the rest of the population. The key question is what are the indicators of the individual water intake occurring in them and how much an increase in the share of this type of consumers affects the height and specificity of water intake from rural water systems [18].

During the period of essential water supply, public water supply facilities shall ensure that the following needs are met:

• population q_u = 15dm^3/res·d;

• food industry and other plants whose functioning is necessary to ensure the survival of the population - individually depending on the needs of the planned volume of production in special conditions (with the separation of plants which source water from their own water supply systems);

• public utility plants with 50% of the normal water demand;

• decontamination and deactivation, including for the following purposes:

a) 45 dm^3/person of the permanent and field sanitary treatment institutions,

b) stationary and field clothing decontamination units in a quantity of 40 dm^3/1kg of disinfected clothes,

c) stationary and field transport decontamination units in a quantity of 200-500 dm^3/1vehicle depending on the vehicle size,

d) stationary and field veterinary posts in the following quantities:

- for a large animal 20÷25 dm^3/1 animal,
- per sheep, goat 12÷15 dm^3/1 animal,
- for pigstock 4-5 dm^3/1 animal,

with higher values to be taken into account for the calculation of the necessary amount of water;

• land and technological facilities - 15% of the water produced at that time.

The amount of water for fire-fighting purposes is determined in accordance with the needs and specifications of the area, according to the regulations of the fire brigade's field commands.

During the period of supply of minimum quantities of water, public water supply facilities shall ensure that the following needs are met:

• population - in the number of 7.5 dm^3/res·d,
• livestock animals as indicated in Fig. 1,

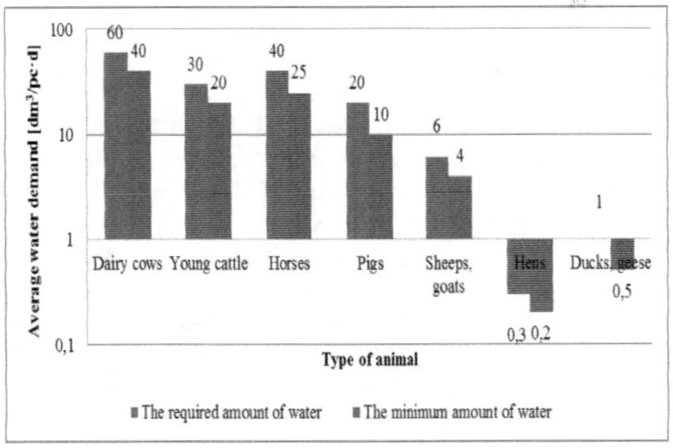

Fig. 1. Average water demand indicators for farm animals during the limited water supply [made on the basis 22].

• sanitary and veterinary treatment points,

• water requirements for the decontamination and deactivation of humans and animals during the minimum amount of water should be set at 50% of the calculated

demand - in accordance with the recommendations concerning the so-called necessary amounts of water.
Generally, all animals should have permanent, unlimited access to the water. Insufficient amount of this constituent limits animal feed intake, which in turn makes production quality worse. The water requirements for livestock are quite varied, depending on the species and age of the animals, as well as the watering system (animal installations with water supply and drinking troughs and no installations). According to the scientific opinion [8-9] and the practice of the zootechnical services, particular attention should be paid to the proposed unit amount of water for young cattle. It seems [19-20], that the so-called minimum quantity of water (q_{min}) for this age group of animals should be increased, from the MSPC's 1995 regulation [6] so far proposed, from q_{min} = 20 dm^3/animal index by min. 50%, i.e. to $q_{min\ rect.}$ = 30 dm^3/animal, where $q_{min\ rect.}$ is rectified minimum quantity of water. This is due, among other things, to the reasons for survival and counteracting the degeneration of this age group of cattle. It also appears that in critical situations – the absolute minimum public water supply - the unit rate may reach a value of q_{min} = 3 dm^3/res·d, but only for a period not exceeding 3 days. After this period, the water supply of q_{min} = 7.5 dm^3/res·d should be increased.

When calculating the required water requirement and minimum water quantity, the coefficient of unevenness of water distribution per day (N_d) should be taken into account in the value of N_d = 1.1. For a limited water supply period, 15% should be taken into account - in addition to the calculated average daily necessary and minimum water requirements. In turn, the approximate values of coefficient of unevenness of daily water distribution are as follows [12,19]:

- N_d = 1.5 for animals: horses, cattle, pigs;
- N_d = 1.3 for other groups of animals, including: goats, sheep, poultry, nutrias, foxes and mink, rabbits.

The average values of the coefficient of unevenness of hourly water decomposition are as follows [10]:

- N_h = 3.0 for all types of animals by watering them from the water supply system in the livestock building;
- N_h = 4.5 by watering animals (horses, cattle, pigs) without water supply system.

According to the studies [20] it was suggested that the so-called integrated indicators of water demand and wastewater discharged were adopted, which are the result of many years of ILRGF (The Institute for Land Reclamation and Grassland Forming) research and trends of the actual decrease of water consumption in rural settlement units covered by the collective water supply. Integrated indicators of the demand for water and the amount of household and economic wastewater from rural units cover the relevant needs of households, small dispersed institutions, public utilities (commune offices, schools, shops, artisanal service points) without taking into account the demand for water for watering garden plots, herbal plots, crop spraying (plant spraying), cooling wastewater, vehicle washing and agricultural equipment (aside from buildings). It has been assumed that the water demand according to the so-called integrated indicators will be 100% equivalent to the amount of sewage generated.

The second assumption is to take into account the value of the coefficient of unevenness of wastewater outflow $N_{d\ sew\ san.}$ reduced by approx. 20% compared to $N_{d\ water}$.

Also the concept was adopted assuming that, for the calculation of water requirement and the amount of generated municipal wastewater from a dedicated business unit located in the village, if less than 30% of the total demand, and the amount of wastewater, the units are not included in the tables of summary (water and sewage).

For such assumptions integrated indicators of water demand and household sewage drainage would shape up as follows [21]:

- municipal settlements and villages with developed services, including recreational ones

$$q_{u\ integr.} = 250\ \mathrm{dm^3/res \cdot d} \qquad N_d = 1.1 \qquad N_h = 1.8 \tag{1}$$

- the other housing classified into public water supply

$$q_{u\ integr.} = 200\ \mathrm{dm^3/res \cdot d} \qquad N_d = 1.1 \qquad N_h = 2.0 \tag{2}$$

- estates qualified for an individual water supply

$$q_{u\ integr.} = 160\ \mathrm{dm^3/res \cdot d} \qquad N_d = 1.2 \qquad N_h = 1.6 \div 2.0 \tag{3}$$

where:
$q_{u\ integr.}$ – unit integrated water demand indicator;
N_d – daily coefficient of unevenness;
N_h – hourly coefficient of unevenness.

In the case of adoption of the extended scope of the demand for water, consumers supplied with communal water systems in rural areas, i.e. in relation to the aforementioned assumptions about the needs of the animal husbandry operation of vehicles and machinery, watering home gardens, values of integrated indicators will increase to amount [20]:

- $q_{u\ integr.}$ = 400 dm^3/res·d for municipal housing with decreasing tendency to approx. $q_{u\ integr.}$ = 300 dm^3/res·d and rural with well-developed services (including recreational villages);
- $q_{u\ integr.}$ = 300 dm^3/res·d for other units of decreasing tendency to approx. $q_{u\ integr.}$ = 250 dm^3/res·d.

The demand for water for the water-absorbent facilities, i.e., those in which water needs exceed $Q_{dayaverage}$ ≥ 50 m^3/d ($Q_{day\ average}$ – average daily water demand) and the livestock farms with the density exceeding 100 LU (LSU, Livestock Unit) is calculated separately and added to the water demand for the individual as calculated from the integrated water demand standards [10,20].

1 LU value is an animal having a body weight of 500 kg or a number of animals with total body weight of 500 kg. It is also important to consider other indicators in terms of conversion techniques, e.g. large manure unit (LMU), which produces 10 tons of manure and a large feed unit (LFU), which means that it consumes 3500

ounces of oats. The conversion factor for LU is commonly used, but the conversion of the LMU and LFU of animals allows you to calculate specific water requirements for specific purposes. The calculations made by the authors make it possible to conclude that the largest daily average water consumption is for large manure unit, and the water requirement for large units is higher for pigs than for cattle.

General water demand (as $Q_{day\ average}$) for the animals in special conditions, according to the author, can be calculated using the following relationship:

$$Q_{day\ average} = a_1 \cdot N_1 + a_2 \cdot N_2 + \ldots + a_n \cdot N_n \text{, (m}^3\text{/d)} \quad (4)$$

where:

a_1, a_2, a_n - individual indicators of water demand for each group of animals, expressed in m^3/piece·d;

$N_1, N_2 \ldots N_n$ - number of animals in each group.

When expressed in terms of water demand indicators for individual animal groups in dm^3/animal·d, formula (4) shall be given as follows:

$$Q_{day\ average} = 10^{-3}\ (a_1{\cdot}N_1 + a_2{\cdot}N_2 + \ldots + a_n \cdot N_n)\text{,(m}^3\text{/d)} \quad (5)$$

$$Q_{day\ max} = Q_{day\ average} \cdot N_d\text{, (m}^3\text{/d)} \quad (6)$$

assuming $N_d = 1.1$.

4 The example of the balance of the average amount of water required for breeding the selected livestock

Calculations of the average amount of water needed for breeding (Table 1) were carried out for selected examples of two groups of animals, i.e. pigs and poultry. The full balance sheet can be carried out taking into account other animal groups: cattle, including cows, sheep and others. Both extensive and intensive animal husbandry was established, with 67% and 33%, respectively. The number of animals was adopted on the basis of 2015 data [22]. Taking into account the total number of animals, we can observe a downward trend in the case of pigs, with the number of 10590.2 thousand in December 2015 and a systematic increase in the number of poultry at the level of 153,177.4 thousand.

Conducted studies and analysis showed that the overall average annual water requirement for livestock under inadequate water supply is around 50% of normal demand. The average amount of water used for breeding pigs and poultry only under conditions of limited water supply is approximately 68% of the water requirement under normal water supply conditions.

In both cases of the conditions of the water supply, the main item in the structure of the overall water demand during the year is feeding and watering animals. For the cases considered, this is 64% in normal conditions and up to 81% in special conditions. Animal health and animal housing makes a total of 33% of the total demand in normal conditions, but in special conditions this position, referring mainly to disinfection and deactivation of the animal rooms is only 14% of the total requirement.

5 Summary

The adoption of specific water demand indicators is based on the objective for which they are adopted. The development of livestock farming depends, among other things, on equipping livestock facilities with water supply and sewage systems, which include watering animals, preparation of fodder, maintenance of animal hygiene and housing.

Table 1. Summary of the average amount of water needed for animal husbandry.

Type of animals	Quantity of animals [mln of pcs]	Average amount of required water									
		Conditions for unlimited water supply									
		Indicators [dm^3/pc·d]					Total in a year [mln m^3]				
		Feeding and watering	Animal hygiene	Premises hygiene	Slaughterhouses	Total[1]	Feeding and watering	Animal hygiene	Premises hygiene	Slaughterhouses	Total
Extensive breeding											
Pigs	7.06	20	2	3	500	25	51.54	5.15	7.73	3.53	67.95
Poultry	102.14	0.5	0.2	0.3	-	1.0	18.64	7.46	11.18	-	37.28
Intensive breeding											
Pigs	3.53	22	3	5	500	30	28.35	3.87	6.44	1.77	40.43
Poultry	51.07	0.6	0.3	0.5	-	1.4	11.18	5.59	9.32	-	26.09
TOTAL 171.75 mln m^3/y											

Conditions for limited water supply									
Type of animals	Quantity of animals [mln of pcs]	Feeding and watering	Hygiene and deactivation of animals and premises	Slaughterhouses	Total/ daily (feeding and watering)	Feeding and watering	Hygiene and deactivation of animals and premises	Slaughterhouses	Total
Extensive breeding									
Pigs	7.06	20	5	500	20	51.54	0.04	3.53	55.11
Poultry	102.14	0.3	0.3	-	0.6	11.18	11.18	-	22.36
Intensive breeding									
Pigs	3.53	20	5	500	20	25.77	0.02	1.77	27.56
Poultry	51.07	0.3	0.3	-	0.6	5.59	5.59	-	11.18
TOTAL 116.21 mln m^3/y									

[1] Total/daily with the exception of the slaughterhouses

From the point of view of agricultural activity of rural farms focused on a specific specialization, in the field of water management, it is important to operate the unit water demand indicators according to specialist literature or on the basis of a technological production project. In order to simplify the calculation of water demand and consumption, it is proposed to take the values of the aggregated indicators for rural areas into account.

Starting from the actual stocking of animals in physical arts, from their conversion into LSU in individual species and from the balance of water in conditions of danger or war, the integrated conversion factors LSU, LMU, LFU for cattle and pigs were determined. The average daily water demand for large manure, feeding and conversion units is:
- for cattle: 122 dm^3/LMU, 121 dm^3/LFU, 113 dm^3/LSU;
- for pigs: 379 dm^3/LMU, 169 dm^3/LFU, 350 dm^3/LSU.

Differences in results are the result of the use of specific conversion factors and result from the animal population. Large pigs have a greater demand for water, as more pigs than individual cattle are needed to achieve one manure, nutrition or conversion unit. Calculations in this area allow to estimate for what purposes there is a need to provide more water. For both cattle and pigs, the largest water demand was demonstrated for a large manure unit. In accordance with the conversion factors used, ultimately the higher water demand for cattle was reached for the LFU and the lowest for the LSU, however, the values are close to each other. In the case of pigs, clearly higher values have been achieved for LMU and LSU, and in relation to them the water demand for a large nutritional piece is more than 2 times smaller.

The issues examined concerning the water requirements for animal husbandry should be continued and extended to include the supply of water for the population in special conditions. It is expected to confirm the fact that there are significant differences in the water demand in the individual conditions and this is due to the current needs of the population. Under normal conditions, the demand is not only for drinking water, but also for hygienic, sanitary, cooking and even recreational purposes. Under special conditions, these quantities should be limited to cover the needs for survival.

The above considerations concern the provision of water in quantitative terms, however, for a full analysis of the issue, the quality requirements of water used in rural areas and agriculture should be taken into account. It is justified to develop an in-depth analysis and assessment of the water quality impact with parameters exceeding the obligatory quality requirements in this respect, ie. toxic contamination, mineral compounds, water pathogenic infections, on the development, health and productivity of farm animals, which will be used for the purpose of developing safety recommendations in this respect. It can be assumed that if the water supply to animals is carried out from the same water supply system, from which water is drawn for the living and economic and drinking needs of the population, there is no need for a separate study of the quality of water used in livestock production. This applies in particular to breeding centres (including rural areas) equipped with water supply systems ensuring adequate water quality.

It is also justified to elaborate the principles of functioning of collective facilities and individual supply of water to villages in special conditions, including technical and design requirements, technical and operational requirements, as well as the rules of organization and functioning of collective and individual rural waterworks.

It is desirable to carry out an analysis of the possibilities of using water from drainage systems to supply drinking water for livestock.

References

1. J. Rak, L. Opychał, GWiTS, **11**, (2009)
2. P. Dohnalik, Z. Jędrzejewski, *Effective operation of water supply - reducing water losses* (LEMTECH Konsulting, Kraków, 2004)
3. Act of 18 July 2001 Water Law (Journal of Laws, 2001, No. 115, item 1229) with later amendments [In Polish]
4. Act of 7 June 2001 on Collective Water Supply and Collective Sewage Disposal (Journal of Laws, 2001, No.72 item 747) with later amendments. [In Polish]
5. Regulation of the Minister of Infrastructure of 14 January 2002 on average standards for water consumption (Journal of Laws, 2002, No. 8, item 70) [In Polish]
6. Regulation of the Minister of Spatial Economy and Construction of 21 September 1995 on the principles of ensuring the functioning of public water supply facilities in special conditions (MP 1995 No. 59 item 663) [In Polish]
7. Regulation of the Council of Ministers of 25 June 2002 on the detailed scope of activities of the Chief of the National Civil Protection, chiefs of civil defense of voivodships, countries and communes (Journal of Laws, 2002, No. 96 item 850) [In Polish]
8. M. Sikorski, J. Simoni, Seminar Materials PZITS, ZP, **7/83**, Sanitary Technique of the Village, 407 (1984)
9. W. Sawicki, *Water and sewage* (PWRiL, Warszawa, 1982) [In Polish]
10. W. Sawicki, *Water in animal production* (PWRiL, Biblioteka Wiadomości IMUZ, No. 40, Falenty, 1972) [In Polish]
11. W. Sawicki, *Technical solutions for the supply of animals in pasture in water* (Instructional materials, Nr 7. IMUZ, Falenty, 1975)
12. S. Drupka, M. Sikorski, *The role of water in the production of essential plant and animal products and the potential to replace certain products in the event of a water shortage* (IMUZ, Falenty, 1999) Technical report not published

13. Regulation of the Council of Ministers of 23 December 1986 on water supply and sewage facilities, as well as charges for water and the introduction of wastewater (Journal of Laws, 1986, No. 47 item 234)
14. R. Błażejewski, A. Waack, *Consumption of tap water in selected villages in the Poznan region* (Scientific and Technical Conference on Water Supply in Cities and Villages, Tom III, 337–347, 1996) [In Polish]
15. P. Bugajski, G. Kaczor, Infrastruct. Ecol. Rural Areas, **2** (2005)
16. J. Pawełek, M. Długosz, Guidelines for calculation of water demand in rural areas in the light of the need for their amendment (Scientific and Technical Conference on Water Supply in Cities and Villages, Tom I, 73–79, 1998)
17. J. Pawełek, S. Satora, Agric. Eng. **8** (2001)
18. J. Pawełek, G. Kaczor, Infrastruct. Ecol. Rural Areas, **2** (2006)
19. M. Sikorski, *The role and importance of water in agriculture and its impact on human and animal health* (ILRGF, Falenty, 1996)
20. W. Sawicki, *Methodology for drawing up regional programs for agriculture and rural water supply. Vol. I. Water supply* (Instructional materials, No. 26, Falenty 1978), *Vol. II. Sanitary sewage system* (Instructional materials, No. 39, Falenty 1980)
21. M. Sikorski, M. *Good farming practices. Spatial solutions of homesteads* (Regional Centre for Advice on Agricultural and Rural Development, Przysiek, 2002)
22. Central Statistical Office of Poland, Farm animals in 2015 (2016)

Hydraulic characteristic of selected installation valves installed on various pipes materials

Krystyna Poręba[1], *Weronika* Kowalczyk[1], *Marcin K.* Widomski[2,*], and *Anna* Musz-Pomorska[2]

[1]Graduate of Lublin University of Technology
[2]Faculty of Environmental Engineering, Lublin University of Technology, Lublin

Abstract. Presented studies covered determination of minor pressure losses and values of minor loss coefficients for two selected installation valves: water control globe valve and angle valve, both DN 15. The tested valves were installed on three pipes, including PP 20x3.4 mm, PEX-Al-PEX 16x2.0 mm and Cu 15x1.0 mm. In order to reflect the real operating conditions of the angle valve, the elastic PVC pipe was used. Our researches were performed on the laboratory installation, for the variable flow rate. The obtained results of laboratory studies showed the clear dependence between minor pressure loss and minor loss coefficients of studied valves, and water flow rate (Reynolds number), degree of valves' closure and, in some cases, manner of valves installation and material of pipes. The greatest values of minor pressure loss coefficients, of mean values relevantly statistically grater, were determined for the tested valves installed on the PEX-Al-PEX pipeline.

1 Introduction

Pressure losses in local resistances in complex domestic installations like hot and cold water supply or heating, may cause a significant portion of the total losses (even 40-60%) [1-4]. Therefore, their exact determination is essential to design the aforementioned installations properly. Nowadays, calculation of minor pressure losses based on the withdrawn Polish Standard PN-M-34034:1976 [5], is additionally complicated by considerable amount of available materials and possible ways of connecting them, as well as diversity of fittings and very close location of the adjacent resistances. For the above reasons, values of minor pressure losses calculated basing on the minor loss coefficients contained in the withdrawn PN-M-34034:1976 [5], are not equal to the real values, reported in technical and scientific publications [3,6-14].

Water control globe valve and angle valve belong to the popular representatives of shutoff fitting widely assembled on domestic installations made of various materials. Values of minor loss coefficient suggested by fittings manufactures and literature reports (including researches) for control globe valve are within the range of 6.8 – 10 [5,15-17] while scientific reports showed values up to 12-20 [18].

Analogically, values of minor loss coefficient for angle valve, suggested by literature, are in the range of 2 – 7 [5,15,16,19]. Additionally, in calculation of minor pressure losses on those two tested valves, keeping in mind the considering resistances generated by couplings used in valve's assembly on a various pipe's materials is needed. Exemplary values of minor loss coefficients for straight pipe connections for various materials vary in range 0.25-1.8 [20-23]. Information about the method of taking into account the influence of the distance between neighboring local resistances on pressure losses are practically inaccessible for designers.

This work contains the results of studies of minor pressure losses and minor loss coefficients for variable degrees of valve closure and different values of flow rate. Researches were made for two shutoff valves, both DN 15, with different construction (globe valve and angle valve), installed on pipes made of three materials: PP, PEX-Al-PEX and Cu.

2 Materials and methods

Water flow resistance's researches presented in this work were performed for two selected installation valves: water control globe valve and washing machine shutoff angle valve, both DN 15. The tested valves, using separable threaded coupler, were installed on three different pipe's materials typically used in domestic installations like hot and cold water supply or heating. The following pipes were used in our researches:

Polypropylene pipe PP-R, 20x3.4 mm, internal diameter 13.2 mm, k=0.007 mm, PN20, 1/10 bar, according to PN-EN ISO 15874-2:2005.

Multilayer, cross-linked polyethylene pipe PEX-Al-PEX with aluminum insert 16x2 mm, internal diameter 12 mm, k=0.007 mm, maximal operational pressure 60°C/10 bar and 95°C/6 bar, according to DIN 4726.

Copper pipe Cu, temper condition R220, 15x1 mm, internal diameter 13 mm, k=0.0015 mm, according to EN 1057.

* Corresponding author: M.Widomski@pollub.pl

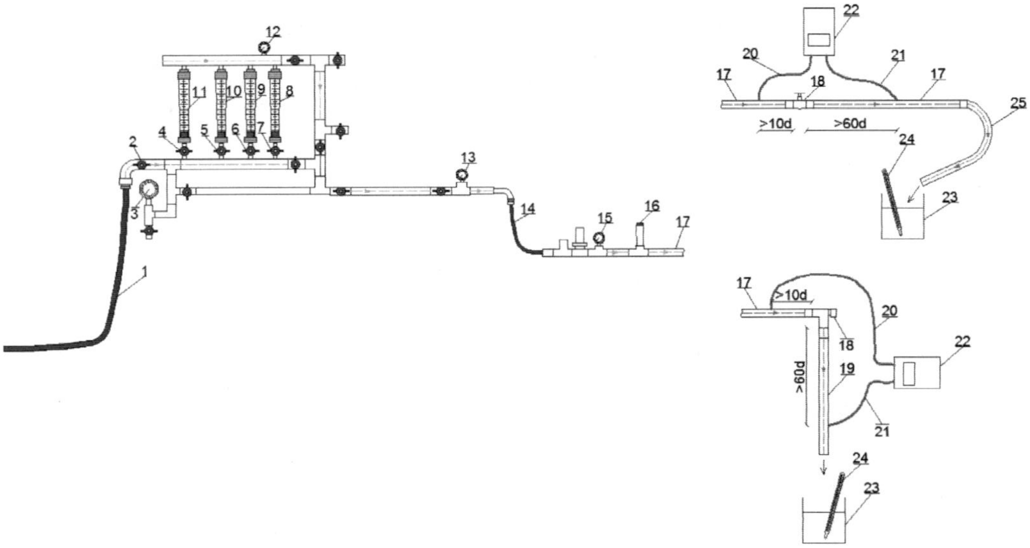

Fig. 1. Scheme of laboratory installation: 1 – elastic inflow pipe, 2 – valve, 3, 12, 13, 15 – manometers, 4, 5, 6 7 – valves, 8, 9, 10, 11 – rotameter, 14 – elastic pipe, 16 – air removal valve, 17 – PP, PEX-Al-PEX, Cu pipe, 18 – studied valve, 19 – PVC pipe, 20, 21 – impulse pipe, 22 – electronic differential manometer, 23 – glass, 24 – thermometer, 25 – elastic outflow pipe.

Additionally, in order to provide angle valve's real working conditions, elastic polyvinyl chloride pipe PVC for cold water 25°C with polymer threaded fittings, internal diameter 10mm, k=0.025 mm was used.

These researches were performed on the laboratory modular installation presented in Fig. 1. Installation consisted of measuring part involving set of interconnected rotameters and interchangeable research part made up of tested valves and the previously described pipes.

Tap water with monitored temperature to feed installation was uptaken from the internal water supply of the building through the reservoir. Set of rotameters used for flow rate measurement was consisted of four rotameters, class 4, measuring range 20 – 200 $dm^3 \cdot h^{-1}$, 60 – 600 $dm^3 \cdot h^{-1}$, 100-1000 $dm^3 \cdot h^{-1}$ and 60 – 600 $dm^3 \cdot h^{-1}$. The rotameters used in the researches were tested by volumetric tank method.

Differences in pressure for selected sections in front and behind the tested valves were measured using the Lutron Electronic electronic differential pressure gauge (range: 0-20 mH_2O, accuracy: 2%) at distances of 10d and 60d [10, 12] respectively, according to PN-EN 1267:2012 [24]. Measurement of flow resistances through the tested valves were performed for variable degrees of valve opening in the flow rate range of 100 – 1560 $dm^3 \cdot h^{-1}$ corresponding to Reynolds number in the range of 2294 – 37202 for globe valve and 100 – 1000 $dm^3 \cdot h^{-1}$, Re=2447-6151 for angle valve. The following degrees of tested valve opening were used: 100%, 75%, 50% and 25%. Application of varying flow rates for both tested representatives of shutoff fitting was due to the difference in the observed minor pressure losses for both cases and the measuring range of the differential pressure gauges. All measuring series, for each flow rate, were performed in at least two repetitions. Determination of minor loss coefficients for tested valves was based on the Bernoulli equation transformation, respectively for each of the calculation schemes. For globe valve:

$$\varsigma = \frac{2g \cdot \Delta h}{v^2} - \lambda \frac{(l_1 + l_2)}{d} \quad (1)$$

For angle valve:

$$\varsigma = \frac{2g}{v_2^2}\left[\Delta h - \frac{v_2^2 - v_1^2}{2g} - \left(\lambda_1 \frac{l_1}{d_1}\cdot\frac{v_1^2}{2g} + \lambda_2 \frac{l_2}{d_2}\cdot\frac{v_2^2}{2g}\right)\right] \quad (2)$$

where:

ς – pressure loss coefficient [-];
Δh – pressure losses read from the differential pressure gauge [mH_2O];
g – gravitational acceleration [$m \cdot s^{-2}$], assumed to be g=9.81 [$m \cdot s^{-2}$];
v_1,v_2 – mean velocity of the fluid flowing in the pipe in front and behind valve, for globe valve $v=v_1=v_2$ [$m \cdot s^{-1}$];
λ_1, λ_2 – Darcy-Weisbach friction coefficient, for globe valve $\lambda = \lambda_1 = \lambda_2$ [-];
l_1,l_2 – pipes lengths in front and behind valve (from the valve to the connection point of the differential pressure gauge) [m];
d_1,d_2 – pipe diameters in front and behind valve, for globe valve $d=d_1=d_2$ [m].

Darcy-Weisbach friction coefficients for each measurement were determined based on the Colebrook-White equation, based on the Reynolds number (Re), the absolute pipe roughness (k) and the internal diameter of the pipe.

Statistical analysis to assess significance of the observed differences between the average values of minor loss coefficients for the same degrees of opening

https://doi.org/10.1051/e3sconf/20185900023

of valves installed on various pipes materials, were based on standard statistical procedures: checking the normal distribution using the Shapiro-Wilk test, then assessing significance of the differences with Fisher ANOVA supplemented by Tukey post-hoc tests and non-parametric Kruskal-Wallis test with multiple comparisons.

3 Results and discussion

Results of presented researches covered determination of minor pressure losses and values of minor loss coefficients in relation to Reynolds number, for tested valves.

Fig. 2 and Fig. 3 present results of minor pressure losses measurement in relation to Reynolds number for two tested valves installed on each of the three used pipe materials. In both cases, there was a clear correlation observed between the value of minor pressure losses for the tested local resistances and change flow geometry caused by closure of the valves. In all tested cases, a significant increase in minor pressure losses for a given Reynolds number was observed with an increase in the degree of valve closure. For fully open globe valve, the pressure losses were in the range of 0.03 – 8 mH_2O for Reynolds number in the range, approx. 2400 – 39000. Closing the valve to 25% of opening resulted in an increase in pressure losses to the range of 0.3 – 17 mH_2O, for Re=2400 – 38000. Similarly, for the angle valve, pressure losses increased from the range of 0.1 – 8.9 mH_2O for full valve opening and Re=2500 – 29000 to the range of 2.0 – 19.5 mH_2O for 25% of valve opening, Re=10 000. Additionally, it can be observed at graphs Δh_m=f(Re) presented in Fig. 2 and Fig. 3, that the highest values of resistances, for each of the degrees of tested valves opening, occurred in case of installing them on multilayer pipes PEX-Al-PEX.

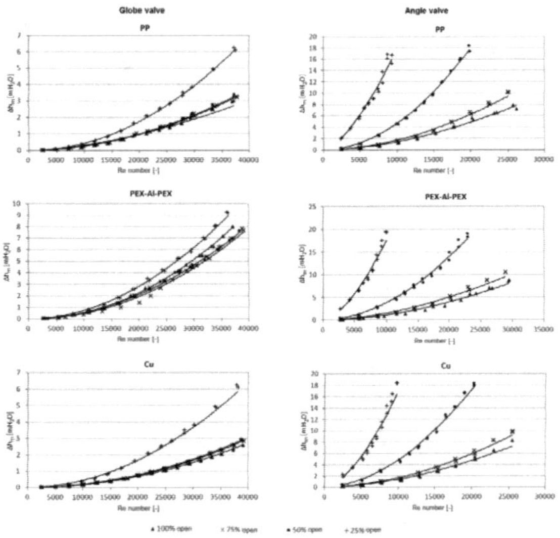

Fig. 2. Scheme of laboratory Minor pressure loss in relation to Reynolds number and degree of valve opening.

Results of calculations of values of minor loss coefficients ς for tested valves together with fittings are presented in Fig. 3. The characteristic values for the individual measurement series are summarized in Tab. 1.

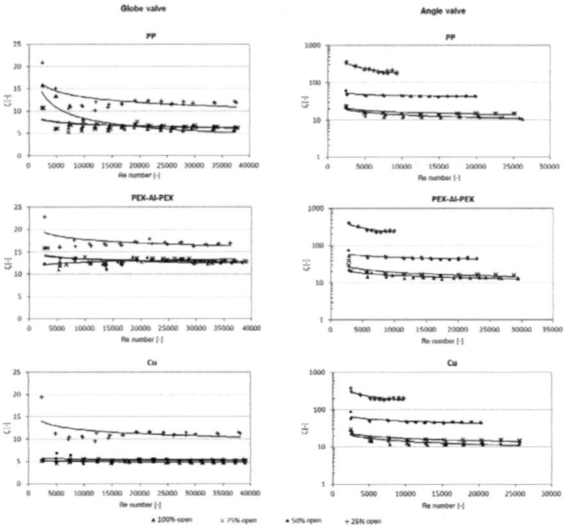

Fig. 3. Coefficients of minor pressure loss in relation to Reynolds number and degree of valve opening.

Results of measurement presented in Fig. 3 showed the dependence between values of minor loss coefficients for tested values together with fittings and Reynolds number. Additionally, there is a noticeable increase in values of coefficients ς in relation to degree of valve closure. For the valve installed on polypropylene pipe with a change in degree of opening from 100 to 25%, the mean value of coefficient ς increased by approx. 70%: from 7.3 to 20.9. On the other hand, for the valve installed on PEX-Al-PEX and Cu pipe there was observed increase of 33% (16.2 – 22.8) and 128% (6.9 – 19.5), respectively. There was also noted dependence of decreasing values of ς =f(Re) for each of opening degree of tested valves for which decrease in the determined value was inversely to the decreased valve opening. In the case of the 25% globe valve opening, the value of minor loss coefficient decreased from about 20.9 to 7.3 for the valve installed on PP pipe, from 22.9 to 16.2 for PEX-Al-PEX pipe and from 19.5 to 6.85 for Cu pipe, for Reynolds number in the range of approx. 2500 – 38000. Similarly, a significant increase in values of minor loss coefficients, depending on the degree of valve opening/closure, was observed for the angle valve installed on the tested pipes. However, both the values of minor loss coefficients and the increment of these values, due to angle valve's construction are considerably higher. In the case of the valve installed on polypropylene pipe, an increase in values of ς was observed, caused by valve closing from full opening to 25%, from accordingly 13.6 to 219.3, which is increase of over 1500%. Similar values were observed for the angle valve installed on the other two pipe materials: an increase from 15 to 163.9 (over

Table 1. Characteristic values of minor pressure loss coefficients for studied valves.

degree of valve opening, %	PP			PEX/Al/PEX			Cu		
	ς min	ς avg	ς max	ς min	ς avg	ς max	ς min	ς avg	ς max
	-	-	-	-	-	-	-	-	-
Globe valve									
100	3.67	6.96	15.83	11.05	13.04	14.17	4.50	4.85	5.45
75	0.54	6.45	10.86	11.99	13.31	19.69	3.40	5.27	10.04
50	0.54	6.59	10.76	11.11	13.02	15.91	4.78	5.43	6.85
25	7.34	11.81	20.89	16.18	17.30	22.83	6.85	11.08	19.48
Angle valve									
100	10.71	13.63	23.25	12.63	14.96	23.32	11.47	13.83	28.27
75	14.08	16.17	24.82	15.00	18.53	40.02	13.80	16.56	29.94
50	41.61	45.11	61.56	42.20	47.86	75.09	42.29	48.40	86.62
25	171.51	219.32	368.97	229.20	263.89	409.12	181.18	208.44	313.87

1660%) and from 13.8 to 208.4 (over 1400%), accordingly for multilayer PEX-Al-PEX pipe and Cu pipe.

There was also a decrease in ς values with increasing flow rate and Reynolds number in consequence, for a given degree of valve opening/closure. For example, for the largest closure (25% valve opening), the value of coefficient ς varied in the following ranges: 369 – 171.5, 409.11- 229.2 and 313.9 – 181.2, respectively for the valve installed on PP, PEX-Al-PEX and Cu pipes.

For both valves, the highest values of minor loss coefficients for a given degree of valve closure were observed for valves installed on PEX-Al-PEX pipe using separable couplers.

The performed statistical analyses showed significance of the observed differences between the average values of minor pressure losses and minor loss coefficients for a given degree of opening of valve installed on PEX-Al-PEX pipe and the values noted for the other two tested pipe materials. Additionally, multiple comparisons and post-hoc tests showed no statistically significant differences between the mean values of flow resistances for fittings installed on PP and Cu pipes. This effect is likely to be explained by approximate shape of flow geometry resultant from realization of socket solder and welded joints. However, in our opinion, this issue requires further researches (e.g. CFD model studies) for the purpose of assessing the influence of flow geometry on velocity spatial distribution and turbulence intensity by the water flow through the tested valve installed on described pipes using separable couplers.

4 Summary

Conducted researches of hydraulic resistances generated by two tested valves installed on three different pipe's materials, showed the dependence between the values of the observed pressure losses and minor loss coefficients and the degree of valve closure. Additionally, the influence of the way of installing valve on a given pipe material on tested flow resistances was observed. Both tested valves were significantly different in their construction, so it is understandable that there are significant differences between the defined minor pressure losses and minor loss coefficients for the same degree of valve opening/closure. Defined minor loss coefficients for the angle valve are few to a dozen times higher than coefficients for the globe valve.

Additionally, an increase of minor loss coefficient for angle valve was observed up to dozen times in the event of changing valve opening from fully open to 25% of opening.

The above results suggest that the sudden increase of generated hydraulic resistances, significantly lowering the pressure behind the discussed valve, makes this valve not to be used as a control valve, as it may adversely affect the operation of domestic appliances connected to the water supply by the tested valve.

The researches also showed that the installation of the tested valves using separable couplers directly affects the value of minor loss coefficients. For each tested degree of valve opening, the highest values of observed pressure losses and determined coefficients ς were noted for valves installed on multilayer PEX-Al-PEX pipe.

The determined mean values of minor loss coefficients for fully open globe valve installed on polypropylene and copper pipe were similar to the literature values suggested in the design guidelines. In the event of valve installation on the multilayer pipe, the values of coefficient ς for the full opening of the globe valve were higher than the suggested values for design and belonged to the range known from scientific publications. The average values of minor loss coefficient for full opening of the tested angle valve in each way of its installing on tested materials far exceeded the literature values.

In our opinion, the design calculations of domestic installations should take into account not only the resistances generated by coupler but also the influence of the resistances located close to each other (set of the valve and couplers).

References

1. Installation system Solter PEX, Technical guide, www.hydrosolar.pl
2. I. Cisowska, A. Kotowski, FCEE **8**, (2006)
3. P. Grajper, J. Smołka, Gaz Woda, 7/8 (2010)
4. K. Weinerowska-Bords, Instal **6**, (2014)
5. Polish Design Standard PN-M-34034:1976
6. T. Siwiec, D. Morawski, G. Karaban, Gaz Woda **2**, (2002)
7. I. Cisowska, A. Kotowski, Gaz Woda **10**, (2004)
8. K. Strzelecka, K. Jeżowiecka-Kabsch, Ochr. Srod. **30**, 2 (2008)
9. K. Strzelecka, K. Jeżowiecka-Kabsch, Ochr. Środ. **32**, 1 (2010)
10. M.K. Widomski, A. Musz, M. Iwanek, Gaz Woda **9**, (2012)
11. F.G. Piechurski, *Water and sewage installations, 3rd Scientific and Technical Conference* (Warsaw-Dębe, 2009)
12. K. Dul, M.K. Widomski, A. Musz-Pomorska, Instal, **11**, 36-39 (2015)
13. N. Gietka, Acta Sci. Pol. Formatio Circumiectus **14**, 1 (2015)
14. M. Kalenik, Rynek instalacyjny **11**, (2013)
15. D.G. Newnan, *Fundamentals of Engineering* (Chicago: Kaplan AEC Education, 2004)
16. B.E. Larock, R.W. Jeppson, G.Z. Watters, *Hydraulics of Pipeline Systems* (Florida: CRC Press, 2000)
17. W. Kołodziejczyk, M. Płuciennik, *Guidelines for the design of central heating installations* (Warsaw: COBRTI INSTAL, 2001)
18. E. Nowakowski, Gaz Woda **4**, 20-23 (2005)
19. A. Górecki, Z. Fedorczyk, J. Płachta, M. Płuciennik, A. Rutkiewicz, W. Stefański, J. Zimmer, *Water, heating and gas installations, for gaseous fuel, made of copper pipes. Application and design guidelines* (Wrocław: Biblioteka Polskiego Centrum promocji Miedzi, 2009)
20. Kan Therm system. Designer's guide and contractors, pl.kan-therm.com
21. Frankische. System alpex Export 16-75mm. Technical catalog. www.viessmann.com
22. TECE. Universal installation system. Technical information, www.tece.pl
23. Systempol. Water installations. Technical reference for indoor installation of cold and hot water, heating systems. www.detalmet.pl
24. Polish Design Standard PN-EN 1267:2012.

Methods of routing and sizing of water supply networks

Paweł Suchorab[1,*], and *Dariusz* Kowalski[1]

[1]Lublin University of Technology, Faculty of Environmental Engineering, Nadbystrzycka 40 B, 20-618 Lublin, Poland

Abstract. One of the first steps in water distribution systems (WDSs) design process is the pipeline routing, defined as geometrical projection of a designed network. The analogical step can also be found in designing of other elements of technical infrastructure, such as energy lines or roads. Moreover, the pipeline routing process influences pipe's sizing and determinates investment, exploitation and maintenance aspects of the whole WDS. Despite its significant meaning, the routing process is still unsatisfactorily supported by mathematical methods and computer aided tools. Therefore, there are continuous researches of more effective pipeline routing and sizing methods. This paper presents a literature review about currently developed methods of network routing and methods of optimal pipeline sizing.

1 Introduction

Duringthe designing of water distribution systems (WDS) one of the first steps is the network routing. The analogical step can also be found in designing of other elements of technical infrastructure, such as energy lines or roads. In water supply sector, the routing process is defined as geometrical projection of a designed network, dependent on settlement unit plan, landform, localisation of water demand points or communication scheme. Network routing is determined by different rules, in accordance to the type of pipelines. Additionally, the routing process is divided on cross-section and plan routing. Plan routing is the initial action in designing process and means routing of pipelines on situation map. On the other hand, cross-section routing is the final step of designing and means the term of distances between the designed water supply pipe and other existing elements of technical infrastructure. Due to the great importance of routing process and its influence on further designing steps such as hydraulic calculations, investing costs or even operational costs, there are continuous researches of more effective way of water supply designing, including the searches of optimal network routing methods [1-4].

The multiplicity of routing methods developed so far, results in various ways of describing water supply networks. One of the most used water supply representation way is the graph theory, where water supply nodes responds to the graph vertexes, while pipes and connection branches respond to graph edges. That kind of representation allowed to applicate, in the process of shaping geometrical structures of networks, many different algorithms for finding the connection route between selected graph vertexes (source and demand points). The other, relatively new, way to describe geometrical shape of network structures is the application of elements of fractal geometry [5]. However, no method developed so far has a universal character and is possible to apply for any urban territories. It is caused by individual character of each design, which is determined by geographical, social, environmental, economic and technical factors [6,7].

The great number of factors influencing the designing process testify its complexity, which causes that the potential designing method should include many options and criteria [8]. Such requirement causes that water supply design task is the complex optimisation problem. So far, the process of plan routing was performed of the basis of knowledge and experience of the designer [9] - in accordance to specified requirements the designer proposed one or a few possible network variants. Such approach corresponded with searching of the local optimal solution. Recently, researches are developing the global optimisation methods fulfilling all designing requirements. The complexity of water supply designing problem causes the necessity of dividing this optimisation task into two steps: searching for optimal network shape (routing) and optimal network parameters (pipes' dimeters, pump curves, etc.) [10]. The following paper presents the literature review of the most important methods of routing and sizing water supply networks and optimisation of the design process.

2 Routing methods

The history of modern water supply systems shows that the network shape always determined its proper

*Corresponding author: p.suchorab@pollub.pl

operation. Among the biggest water supply network of XIX century, many were designed as a structure of one or a few main loops of the greatest diameters with small branches [11]. It is worth to notice that in former designs the structure of water supply pipes was connected into loops, even if the calculations were made for branched structures. That approach proved, that even without the mathematical substantiation, the importance of water delivery certainty was taken into account. Finally, the Cross method [12] significantly influenced the calculations of looped networks, encouraging the designers to applicate the closed water supply networks.

The other major impact on pipelines routing was the idea of describing network by graph theory. It allows to applicate the least cost path algorithms (LCPA). The least cost path algorithm in accordance to water supply networks, most often means the shortest paths because the edge weight are lengths of particular pipes. The most important LCPA are: Dijkstra [13], Bellman-Ford [14,15], Floyd-Warshall [16,17] and Johnson [18]. One of the most popular is Dijkstra algorithm (DA) developed to find the shortest path from a single source in a graph of non-negative edges weights. For over 50 years since publishing, the DA was frequently modified and optimised [19-24]. One of the problems in application of LCPA is the fact, that these algorithms analyse the single connection between neighbouring nodes. It causes difficulties in representation of technical object such us bridges or tunnels, which do not connect neighbouring nodes but in practice are possible way to significantly shorten the path [25]. The other limitation of LCA application is the fact that the shortest path between two nodes responds to a single water pipe. In order to route the whole network, it was needed to applicate the selected algorithm many times repeatedly. Such approach resulted in significant computer processor overload and considerable calculation time.

Despite the great progress in network routing thanks to the graph theory, no universal method has been developed so far. The possible reason of that state was the assuming that the edge weight is the distance between nodes. The calculated on that basis connection paths were therefore the shortest paths. However, the best (optimal) connections between nodes is not always the shortest path [7], which deny the paradigm that 'shorter is better'. Such approach face the researchers with the task to find the parameter which could be decisive in choosing the connection path. One of the possibilities is the lower path resistance [26], calculated in geographic information system (GIS) environment. As the result, the designer receives information not only about the distance but also about time of delivery or reliability factor. Therefore, the final decision about the selected path is often supported by multi-criteria decision analysis (MCDA), according to which the selected solution should fulfil all requirements [27].

Another method of network routing is based on elements of fractal geometry [5], where every network can be presented as dendric structures. In the developed method the iterative duplication and rotation of a basic section is used to shape the geometrical structure of a network. Similar method is also presented in [7], where in addition to the total distance also the total rotation angle is analysed. Basing on analysed methods, it can be said that despite its significant meaning, the routing process is still unsatisfactorily supported by mathematical methods and computer aided tools.

3 Sizing methods and water supply networks optimisation

Proper dimension sizing determines the total investing and operating costs of water supply network, which make pipe sizing an optimisation task. Therefore, the problem of optimal designing the water supply network is searching for global minimum cost dependent on selected diameters [27]. Such approach has several primary assumptions such as: constant geometrical network shape and defined demands in nodes [9]. While the optimal designing of branched networks is well presented in the literature, there is only few looped structures optimal designing methods [10], because of the unknown direction of water flow [28], nonlinear dependence between flow and pressure losses and presence of discrete decisive variables [29]. In addition, in looped networks, the process of pipe sizing is more dependent on the network's shape [10]. In order to optimally design looped network, it is required to optimally route pipelines. However, it is not possible to optimally route network pipes without the knowledge of pipe sizing and vice versa. Therefore, the optimisation of looped water supply networks is an NP-hard combinatorial problem (NPH) [30].

One of the methods of solving NPH problem is linear programming (LP). In paper [31], the successive linear programming gradient (LPG) was used to solve the problem. To more complex water supply structures also the nonlinear programming (NLP) and dynamic programming (DP) is used to find the solution including several required parameters. However, the mathematical programming can be applied only for finding the local optimum, which is not fully satisfactory in accordance to complex looped water supply structures [32]. To find global optimal solution the heuristic methods are used. The heuristic methods allows to find the solution which in acceptable level fulfil the requirements of the optimal solution, but not always is the optimal solution. While traditional methods are impossible or impractical, the heuristic approach allows to solve the NPH problems fast and easy, thanks to available data analysis and predictions of forthcoming phenomenon basing on logical combinations and iterative improvements of developed solution. The example of heuristic method for finding global optimum is the Branch and Bound (BB) algorithm [33].

The more advanced methods are metaheuristics technics, which allows to find the final solution in indirect way – the applied algorithm does not find the precise solution but indicate the heuristic algorithm which will find the solution fulfilling the requirements of optimal solution in an acceptable level. The metaheuristic approach is most common used when the input data is incomplete or inaccurate. The basis of metaheuristic technics are analogies from real world – the rules of natural phenomenon. One of the most popular metaheuristic algorithm is the genetic algorithm (GA) [34]. Both GA and other evolutionary algorithms (EA) found its application in water supply sector: simulated annealing (SA) [35], harmony search (HS) [36], Tabu Search (TS) [37], ant colony optimization (ACO) [7], shuffle frog leap algorithm (SFLA) [38], particle swarm optimization (PSO) [39] orsoccer league competition (SLC) [40]. However, despite many developed methods there is still no universal algorithm ensuring the satisfactory results and effective optimisation process for water supply networks of a real size [11].

4 Conclusions

The necessity of improving WDS reliability enforced delivering water to every consumer with at least two possible paths, which was obtained by connecting branched networks into looped structures. Automatically, it increased the investment and operational costs of water supply networks and therefore in recent years the significant development of optimisation technics and tools can be observed.

Basing on literature review, it can be said that pipeline routing still can be treated as a separate design step but it is mainly used in transmission pipelines routing. For this purpose, the GIS environment is used most often, together with design supporting tools such as MCDA. Network routing, in dependence on selected describing network method, in most often combined with pipe sizing. So, nowadays the most popular are hybrid optimisation methods, which parallel develop the network structure and pipe diameters.

However, despite many developed optimisation methods, there is still no one universal tool, which could be used in any real-size water supply network. Therefore, it is reasonable to continue the researches on such method, which could be use not only in references conditions but also in real water supply networks with multiple water sources.

This article was founded by the statutory activity of the Faculty of Environmental Engineering, Lublin University of Technology.

References

1. M. Cisty, Z. Bajtek, SJCE **16**, 4 (2008) www.svf.stuba.sk/buxus/docs/sjce/2008/2008_4/file2.pdf.
2. G.M. Goncalves, M.V. Paro, Ann. Oper. Res. **94**, (2000)
3. G.M. Goncalves, L. Gouveia, M.V. Pato, Ann. Oper. Res. **219**, (2014)
4. E.J. Lee, D.L. Freyberg, C.S. Criddle, Environ. Model. Softw. **84**, (2016)
5. D. Kowalski, B. Kowalska, P. Suchorab, WIT Trans. Built Environ. **139**, (2011)
6. M. Nussbaum, Right of Way, January-February, (2012)
7. D. Beaza, C. Ihle, J. Ortiz, J. Clean. Prod. **144**, (2017)
8. J.R. Rak, B. Kucharski, Environ. Prot. Eng. **35,** 2 (2009) epe.pwr.wroc.pl/2009/Rak_2-2009a.pdf
9. A.R. Simpson, G.C. Dandy, L.J. Murphy, J. Water Resour. Plann. Manage. **120**, 4 (1994)
10. G. Razei, M.H. Afshar, M. Rohani, Adv. Eng. Softw. **70**, (2014)
11. E. Todini, Urban Water J. **2**, (2000)
12. H. Cross, *Analysis of flow in networks of conduits or conductors* (Engineering Experiment Station, 286, 1936)
13. E.W. Dijkstra, Numer Math **1**, (1959)
14. R. Bellman, Quart. Appl. Math **16**, (1958)
15. L.R. Ford, *Network Flow Theory* (Santa Monica: RAND Corporation, Paper P-923, 1956)
16. R.W. Floyd, Commun. ACM **5**, 6 (1962)
17. S. Warshall, JACM **9**, 1 (1962)
18. D.B. Johnson, JACM **24**, 1 (1977)
19. J. Mistra, IPL **77**, (2001)
20. D. Cantone, S. Faro, ENDM **17**, (2004)
21. M.H. Xu, Y.Q. Liu, Q.L. Huang, Y.X. Zhang, G.F. Luan, Appl. Math. Comput .**185**, (2007)
22. J.B. Orlin, K. Madduri, K. Subramani, M. Williamson, J. Discrete Algorithms **8**, (2010)
23. W. Shu-Xi, Procedia Eng. **29**, (2012)
24. Y. Dinitz, R. Itzhak, J. Discrete Algorithms **42**, (2017)
25. Y. Chaoqing, J. Lee, M.J. Manuro-Stasiuk, Int. J. Geogr. Inf. Sci. **17**, 4 (2003)
26. M. Herrera, E. Abraham, I. Stoianov, Procedia Eng. **119**, (2015)
27. C. Bragali, C.D'Ambrosio, J. Lee, A. Lodi,P. Toth, *An MINLP Solution Method for a Water Network Problem*. (in: Azar. Y., Erlebach T. (eds) Algorithms – ESA 696-707 2006)
28. J. Reca, J. Martinez, Water Resour. Res. **42**, (2006)
29. R. Uma, IJSR **5**, 11 (2016)
30. I. Gupta, J.K. Bassin, A. Gupta, P. Khanna, Environ. Model. Softw. **8**, (1993)
31. E. Alperovits, U. Shamir, Water Resour. Res. **13**, 6 (1977)
32. H.D. Sherali, S. Subramanian, G.V. Loganathan, J. Global Optim. **19**, (2001)
33. A.H. Land, A.G. Doig, Econometrica **28**, 3 (1960)
34. J.H. Holland, *Adaptation in Natural and Artificial Systems* (Cambrige, MIT Press, 1975)

35. S. Kirkpatrick, C.D. Gelatt, M.P. Vecchi, Science, **220**, 4598 (1983)
36. Z.W. Geem, J.H. Kim, G.V. Loganathan, Simulation **76**, 2 (2001)
37. A. Fanni, S. Liberatore, G.M. Sechi, M. Soro, P. Zuddas, Comput. Tools Model. Optim. Simul. **12**, (2000)
38. M.M. Eusuff, K.E. Lansey, J. Water Resour. Plann. Manage. **129**, 3 (2003)
39. I. Montalvo, J. Izaquierdo, R. Perez, M.M. Tung, ORCS **56**, (2008)
40. N. Moosavian, B.K. Roodsari, Swarm. Evol. Comput. **17**, (2014)

The use of mathematical modelling to analyse the process of mixing waters from different sources in water supply system

Agnieszka Szuster – Janiaczyk[1,*], *Rafał* Brodziak[1], and *Jędrzej* Bylka[1]

[1]Institute of Environmental Engineering, Poznań University of Technology, Berdychowo 4, 60-965 Poznań

Abstract. One of the processes that significantly determines the quality of water to consumers is the process of mixing water from different sources in the water mains. Put to the network two or more chemically and biologically stable waters may result in the formation of water that will be deprived of these features. This article presents the german guidelines for analysing water quality for mixing waters from different sources, in various proportions. Then performed an analysis of utility the mathematical models, including quality criteria, for use in network control. An IT tool has been developed to manage selected water quality processes using mathematical modeling. The basis for implementing the tool was a network model created in Epanet integrated with the Matlab.

1 Introduction

The quality of tap water depends on many processes during its intake, treatment and distribution. From many factors influencing water quality in supply network, hydraulic parameters - such as: pressure gradients, flow velocities, water age - are particularly important. Parameters of water pumped into network such as chemical and biological stability, are also important. Process management can be understood as all activities directed for fulfilling the aims and objectives of a company. One of the processes, which influence the tap water quality is mixing of the water from different sources. This process occurs commonly in the water distribution system and contributes to lots of problems in the network [1-3]. The complexity of this problem results from the fact that the quality of water after mixing is a result of both - the parameters of water quality before mixing and the mixing proportion. To analyse the quality of water before mixing, the German DVGW guideline[4] can be used.

Ensuring good hydraulic parameters is possible due to using the tools for water distribution system control. The main aim of network control is to determine and correct the settings of the actuators. Hydraulic modelling tools can be used for supporting this process. Hydraulic modelling of water distribution system consists of solving mass and energy continuity equations for different sets of boundary conditions.

Practically, equations systems are solved using numerical methods (eg. gradient methods) in computer software, e. g. EPANET [5,6]. Using the results from hydraulic modelling calculations, it is possible to calculate the quality of water after mixing. The process of mixing water in the nodes, where it is assumed that water flows from two or more sources and is completely mixed, shall be described by the mass balance in the node represented by the equation (1):

$$C_{oj} = \frac{\sum_{=} Q_{ij} C_{D_i} + U_j}{\sum_{i=D_j} Q_{ij}} \quad (1)$$

where: C_{oj} – concentration of a substance in water leaving the junction j, g/m^3, O_j – for links, in which water outflow from junction *j*, *Dj* – for links, in which water inflow to junction, Q_{ij} – flux of water inflow to junction *j* from *i*'th pipes m^3/s, C_{Di} – concentration in water inflow to *i*'th links g/m^3, U_j- concentration in *j*'th junction g/m^3[6-7].

Mixing of water in storage points (tanks) is calculated by the weight balance. In case where water inflows to the tank it is assumed that it mixes fully. It can be described by the equation (2):

$$\frac{dC_k}{dt} = \frac{Q_{ik}}{V_k}(C_{Di}(t) - C_k) + \theta(C_k) \quad (2)$$

where: C_k – concentration in *k*'th tanks, g/m^3, $C_{Di(t)}$– concentration in *i*'th pipe from which water inflow to tanks in times *t*, g/m^3, Q_{ik}– flux of water inflow to *k*'th tank form *i*'th links, m^3/s, V_k – volume of *k*'th tank , m^3, $\theta(C_k)$ – reaction factor, g/m^3·s [6-7].

Described equations are the basis for modelling water quality in EPANET software that were used in the developed environment. The main task of the developed tools is to calculate the proportion of water mixing in each node (which is not possible directly in EPANET software). Calculated proportions provide the basis for water quality assessment in the context of water mixing. The article also describes standards for assessing the effects of mixing water on its quality. Developed tools can be used to determine control scenarios which minimize the negative effects of water mixing.

*Corresponding author: agnieszka.szuster-janiaczyk@put.poznan.pl

2 Modelling of water quality during its mixing in distribution system

The basis for modelling changes in water quality parameters in water supply networks are data on pressure and flow in pipelines. Often as a basis for analysis of water quality changes in the network is also used to analyze the water age in the system. Quality modelling process itself is a very complex task due to the multitude of individual processes, which must be taken into account in mathematical equations. Among the equations, one must find: (1) equations describing the reaction kinetics model, taking into account their order, (2) equations describing bulk and wall reactions, (3) equations describing fast and slow reactions in system. At present, many researchers are making efforts to generate equations that might have been applied for water quality models like EPANET- MSX, to provide an adequate reflection of processes that shape water quality in the system. So far, the most work is on models describing disinfection processes in the network - chlorine decay model and by-products formation. Single works deal with other issues for example arsenate transport in network. However, the utility and universality of many of the proposed equations have not yet been confirmed in water supply modelling [8-9].

An important issue in analyzing changes in water quality in the network is the analysis of the influence of the mixing of waters from different sources on tap water quality. Numerous literature data [1-3] state that such mixing, as well as replacing one water source with another, might result in deterioration of water quality in the system. For a long time, as the main reason for problems with maintaining the quality of water mixed in the water supply system, the lack of carbon balance was considered [10]. At present, the analysis of water quality changes during its transport, focuses on the analysis of the anti-corrosion protective layers, formed on the surfaces of materials, that come into contact with water. In cases where water is supplied to the system from more than one source, the process of forming this layer is disturbed, which may result in intensification of corrosion processes occurring in the network and deterioration of water quality [10]. This issue has been analyzed in detail in the German Guideline DVGW W216, concerning of supplying to the network the water of variable quality from one source or waters from two and more sources and an attempt to model a process of mixing waters from different sources can also be found in EPANET-MSX [4].

The guideline W216 [4] indicates the parameters (temperature, oxygen, pH, mineral acidity to pH = 4,3, chlorides, sulphates, nitrates, phosphates, dissolved organic carbon, nitrogen, potassium, calcium, magnesium), and the possibility of variations in their concentrations for two and more different waters, to guarantee good water quality in the network. Analysis of two or more waters for their stability during mixing should take into account three main criteria: corrosive-chemistry criterion according to W216, calcium carbonate solubility assessment, and a microbiological criterion. There should be no conclusions about the stability of mixing waters only on the basis of one of these three mentioned criterions. In the case of the microbiological criterion, there are no binding stability-value that would indicate that two waters after mixing in the network won't generate microbial problems. The only way is to perform a detailed evaluation of microbiological regrowth potential for analyzed system. In the case of mixing waters from different sources, depending on the quality of the water before mixing and the mixing ratio, water of stable or temporally variable quality can be achieved. If for each of the analyzed in W 216 parameters, in each mixing ratio, we obtain stable water, it can be mixed in any proportion, if not, the ratios of mixing need to be establish to maintain stability of water during its distribution. In the case of mixing the three waters containing the S component at the starting concentration of S_a, S_b, S_c, with their share in the total volume of mixed water respectively x_a, x_b, x_c, then the concentration of this component in mixed water will be (3):

$$S_m = x_a S_a + x_b S_b + x_c S_c \tag{3}$$

According to the equation (3), the composition of water mixed in case of anions, cations, dissolved inorganic carbon, acidity, alkalinity, ionic strength and temperature can be determined. For substances that react during the mixing process, i.e. hydrogen ions, acidic carbonate ions, carbonate ions, carbon dioxide, their concentrations after mixing must be determined only after taking into account the chemical equilibrium.

Also in the EPANET-MSX, creators declare the possibility of water quality modelling in situations, when waters from different sources and different quality are delivered to the water supply network. However, the literature shows, that verification of the model's performance in this task is usually based on the reaction of chlorine decay in the network, i.e. reaction which are relatively simple to modelling, compared to other reactions in the water supply network. In addition to this, the main and significant simplification of this analysis is that in EPANET-MSX it is assumed that the reaction rate coefficient for mixed water is the weighted average of reaction rate coefficients for two mixing waters. That's why, the results of the analyzes are often unsatisfactory and show significant differences with actual chlorine concentrations measured in the distribution system. For example, Grayman with the team [9] has attempted to use the EPANET-MSX capability to determine the chlorine bulk decay coefficient to individual pipes for networks.This method determines the percentage of water from each source of water reaching the individual pipes in the individual time steps, as illustrated in equation (4):

$$Kb(i,t) = [KB1 \cdot TR1(i,t)] + [KB2 \cdot TR2(i,t)] \tag{4}$$

where: $Kb(i,t)$ – bulk coefficient for pipe i at time t, $KB1$ – bulk coefficient for water from source 1, $KB2$ – bulk coefficient for water from source 2, $TR1(i,t)$ – fraction of water in pipe i from source 1 at time t, $TR2(i,t)$ – fraction of water in pipe i from source 2 at time$t$, and TR1 (i, t) + TR2 (i, t) = 1 [9].

The method can be generalized for any number of water sources, but the results reached from network have prompted the authors to draw the conclusion that further work is needed to adapt the equations for modelling mixing of waters from several sources [9].

3 Developed environment

EPANET allows the user to perform three types of basic qualitative analyzes, such as: water age, tracking of the spread of the substance and calculation of changes in chemical concentration. TRACE functions can be used in the water mixing analyzes to determine how much water comes from a selected water source, at any node of the network. The EPANET GUI limitation is that this feature allows you to check only one source on each node for one simulation run. This means that checking the mixing ratios of water from more sources can be problematic. To define ratios of water from each source, the user needs to perform separate simulations, in which mixing water ratio for one reservoir can be read. Applying this solution can be time-consuming and will require running many separate simulations controlled by the user. From the point of view of the results analysis, it is also important to be able to present data in a variety of forms, such as graphical charts of the mixing ratios values. The possibilities are very limited in EPANET GUI.The purpose of the work was to prepare an environment that would allow the user to analyze the mixing of water in the network in one simulation run, to present results in graphical form and to compare the mixing parameters for various variants of system operation. For the purpose of analyzing the mixing ratios of all power sources in the individual nodes of the water supply model, an environment integrating EPANET functions with MATLAB was used.

This was possible with the available source code and the Programmer's Toolkit [12], including the DLL library.

They were used to create classes that call EPANET functions from MATLAB [11]. The water supply model for the analyzed sample was created in the EPANET GUI from where it was exported to the INP file. The model in this text file was read in a Matlab environment where multiple simulations with automatically changing parameters were performed and the results were saved for further processed (Fig. 1).

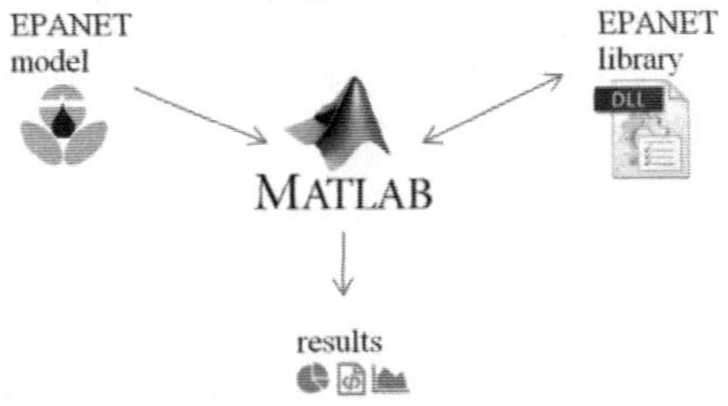

Fig. 1. The concept of integration tools in the developed simulation environment [11].

The set Quality Type function was used to perform the analysis. Iteratively, using it, the modelling mode was specified with an indication of each tracked source. For a given simulation time, water quality was measured at each node, saving the results for each hour.

4 Case study

The prepared environment allows to check the mixing ratio of water from different sources in each node. In the analyzed example, modelled a simple annular water-supply system supplied from three sources. Fig. 2 shows a diagram of the network and the results of simulations in which the determined amount of mixing water in the nodes presenting a pie charts, the ratio of water from each source.

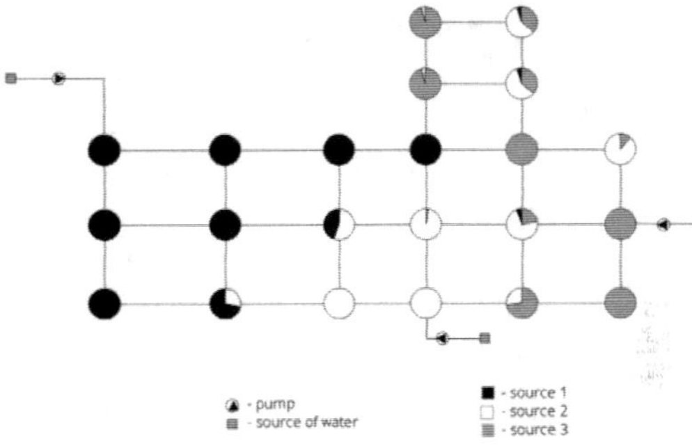

Fig. 2. Mixing ratios of water in nodes for three sources.

In the prepared environment it is possible to calculate the water mixing value for each node over time. This value can vary depending on the adopted water change demand pattern or the pump control schedule.

Fig. 3 shows the variation in water volume ratio from each source for one selected node. Note that the EPANET program assumes that at time 0 (at the beginning of the calculation) the network is filled with water, for which the parameters are assumed as if the water did not come from any source. It follows that the results of calculations become reliable only after a certain time, after which the water exchange in the whole water supply network will take place.

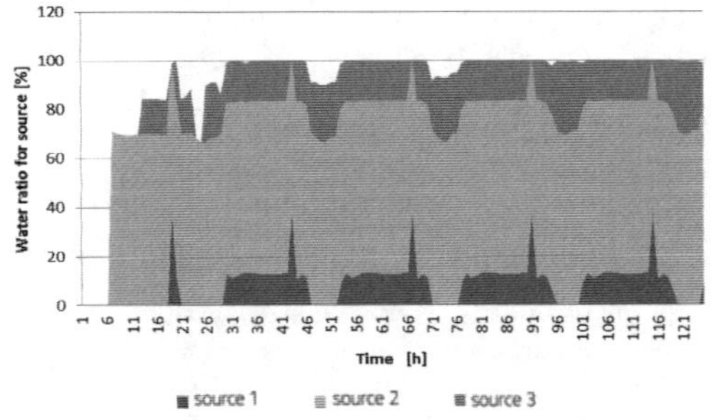

Fig. 3. The proportion of water from various sources in the selected node-I supply pattern.

For example, for the analyzed network, reliable results can only be considered after the second day. After this time, the cyclicality in the repetition of the results is noted. It should be noted that the mixing ratio of water in the network is variable in time, e.g. during the night hours only water from two sources is mixed. The situation resulting from this analysis may be the basis for deciding to change the pump control algorithms.

In the prepared environment you can see how the water mixing ratio in the node changes due to other pump control. An exemplary case in which a reduced rotational speed of the pump motor, pumps water from a source 3 shown in Fig. 4.

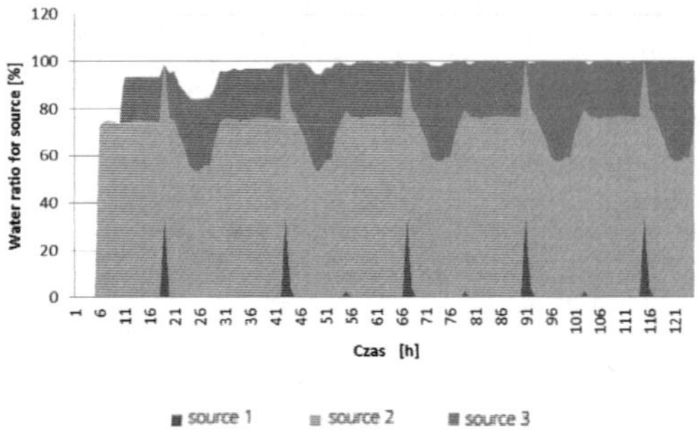

Fig. 4. The proportion of water from various sources in the selected node - II supply pattern.

In the developed environment it is possible to test many - freely prepared scenarios. The use of this type of solution can be helpful in the selection of optimal scenarios and to locate zones where most of the time will not reach the water supply from various sources, which can help reduce the negative effects of mixing water of different quality.

4 Conclusions

A review of literature and the feature's library of EPANET-MSX shows that there is still no possibility to fully model of water quality in network, especially when the waters from two or more different sources are mixed in system. The limitation is that, (1) there are still no enough resources of equations, describing network reactions, (2) there are problems in determining the rate coefficient of reaction (especially for wall reactions for pipes, which are located in the mixing zone), and (3) limited effectiveness of hydraulic models in variable pressures.

The paper presents an analysis of mixing water process, in situation when waters from two or more sources supply the network, based on the German DVGW guidelines and EPANET program. The proposed configuration of analyzes enables forecasting changes in water quality in the network resulting from its mixing process, among other things based on the nomograms contained in the guidelines. The information tool provided facilitates the control of the water supply network in the case, when the quality of the water, which is mixed in the network and the proportion of this mixing is an important criterion of decisive.

References

1. J. Bylka, A. Szuster-Janiaczyk, Water Sci. Technol. Water Supply, (2016)
2. J. Chen, H. Qian, P. Li, Water **5**, (2013)
3. F. Montiel, C. Nguyen, Water Sci. Technol. **45**, 4/5 (2002)
4. Technische Regel: Versorgung mit unterschiedlichen Trinkwassern, DVGW, W 216, Bonn (2004)
5. O. US EPA, "EPANET." [Online]. Available: https://www.epa.gov/water-research/epanet. [Accessed: 29-Jun-2016]
6. T. M. Walski, D.V. Chase, D.A. Savic, W. Grayman, S. Beckwith, E. Koelle, *Advanced water distribution modeling and management* (Bentey Institute Press, Exton, Pensylvania USA, 2007)
7. F. Shang, J. Uber, L.A. Rossman, *EPANET Multi-Spacies Extension User's Manual* (Environmental Protection Agency, PA 600/S-07/021, Cincinnati, OH, January, 2008)
8. I. Fisher, G. Kastl, A. Sathasivan, Water Reas. **45**, (2011)
9. W.M. Grayman, S. Ksihrsager, M. Rivera-Sustache, M. Ginsberg, JWMM, (2012)
10. A. Bauer, G. Dietze, W. Müller, K.J. Soiné, Weideling D. *Handbuch für Wassermeister. Wissenswertes für den Betrieb von Wasserversorgungsanlagen*, (DVGW, Oldenbourg Verlag, 1998)
11. R. Brodziak, J. Bylka, A. Urbaniak, P. Zakrzewski *Modelling And Simulation - The European Simulation And Modelling Conference* (*ESM '2016*, J. Évora-Gómez and J. Juan Hernández-Cabrera, Eds. Las Palmas, Gran Canaria, Spain: EUROSIS-ETI, 2016)
12. "KIOS-Research/EPANET-Matlab-Toolkit,"*GitHub*. [Online]. Available: https://github.com/KIOS-Research/EPANET-Matlab-Toolkit. [Accessed: 29-Jan-2015].

Smart specialization as a factor stimulating innovative development of water and wastewater economy

Joanna Machnik-Słomka[1,*]

[1]Silesian University of Technology, Faculty of Organization and Management, Poland

Abstract. The article discusses the issue of smart specializations treated as a significant mechanism stimulating innovative development of water and wastewater economy. The article shows the links between the developing of smart specializations and the possibility of development innovative, research and development of the economy. The aim of the article is to present the meaning of the concept of smart specialization in the development of innovative water and wastewater economy at the national level and on the example of the Silesian Region. To achieve the goal, the theoretical assumptions of the concept of smart specialization were also presented. In particular, the article emphasizes the need for cooperation, establishing partnerships for the development of smart specializations that contribute to solving problems and increasing the competitiveness of EU countries and regions.

1 Introduction

Water and sewage management is a fixed element in development of civilisations and one of the key factors of socioeconomic development affecting the quality of human life [1]. This management has its specific conditions. On the one hand, the entities performing tasks related to water are fully entitled participants of the competitive market, one the other - one must consider the fact that water "is not a commercial product like any other, but rather an inherited good that must be protected, defended and treated as such"[1]. Thus, taking actions requires consideration of intentions and strategic goals relevant not only for the economy, but also for the society and environment [2].

A response to the challenges and issues related to water and sewage management are implemented EU strategic programmes and environmental standards. They are an impulse for development and implementation of innovations in this area [3,4]. Supporting the development of innovative solutions to deal with water challenges uptake brings significant economic opportunities in a rapidly growing world market, in which many European companies are active and where there is strong potential for job creation [5]. Adapting to EU standards and improving the situation in the sector by implementation of innovations and new technologies requires a comprehensive approach and cooperation of various stakeholders at different levels [6]. Although water is predominantly a local issue, water problems are increasingly globalised, requiring focus at a range of scales, from local responses to global strategies [5].

Opposed to these challenges was the strategy implemented at the EU level related to the concept of smart specialization, which is the implementation element of the European innovation policy. The concept of smart specialisation is oriented at achieving and maintaining competitive advantage of EU countries and regions in the global economy owing to the concentration of knowledge and focusing it on a limited number of priority actions [7]. According to the concept, regions and countries that combine innovation with their unique strengths and capabilities of the economy based on the infrastructure and knowledge bases as well as the industry structure have a greater chance of success [7].

Water technologies fall within the range of national, and frequently regional, smart specialisations in Poland. The subject of the article is discussion of the essence and significance of smart specialisations for development of the water and sewage economy. The issue of smart specialisations treated as a significant mechanism stimulating innovative development of water and sewage economy is referred to the analysis at the national and regional level based on the case study regarding the region of Silesia.

2 The concept of smart specialization and its importance in the innovative development of water and wastewater economy

The concept of smart specialisation is not completely new. It is but an improvement to the existing structural fund planning methodology. It is based on regional experiences in the area of supporting innovative strategies and on the leading economic thought on the part of leading international institutions, for instance the World Bank, or the OECD [8]. The concept of smart specialization is, however, a generation of innovation

* Corresponding author: joanna.machnik-slomka@polsl.pl

policy that goes beyond the general creation of innovation potential and traditional investments in research and technology.

The concept of smart specialisations is the outcome of the work of the K4G: Knowledge for Growth Expert Group appointed in 2005. Assumptions regarding the regional specialisation idea, co-authored by, inter alia, D.P. Foray [9], were presented in 2008 in the working documents of the Group as well as in a report containing recommendations regarding the functioning of the European Research Area [10].

The concept was primarily popularised with the Communication from the European Commission titled Europe 2020: A strategy for smart, sustainable and inclusive growth [11]. Smart specialisation remains closely related to one of the three main priorities of the Europe 2020 – Smart development based on knowledge and innovation strategy. It is also a pillar of one of the seven flagship initiatives known under the name of 'Innovation Union', falling within the framework of the Europe 2020 strategy. Creators of smart specialisation name four key principles on which the concept is based (quadruple 'C'): tough choices and critical mass, competitive advantage, connectivity and clusters, and collaborative leadership [12].

Smart specialisation means identification of unique characteristics and assets of each country and region, highlighting of the competitive advantage of each region, and concentrating regional partners and resources around an accomplishment-oriented vision of their future [8]. It also means strengthening of regional innovation systems, maximising of knowledge transfers, and dissemination of innovation benefits throughout the entire regional economy [8].

In practical terms, of crucial importance was the Guide to Research and Innovation Strategies for Smart Specialisations (RIS3), published in May 2012 and developed with the use of a smart specialisation platform. The Guide contained detailed instructions for regions and Member States with respect to methods of development and implementation of research and innovation strategies for smart specialisations. Development of the 'Research and Innovation Strategies for Smart Specialisations (RIS 3) is designed to allow the use of EU structural funds more efficiently and to increase the synergy between different EU, national and regional policies as well as public and private investments [8]. RIS 3 is a key part of the proposed reform of the EU cohesion policy that supports thematic concentration and strengthens strategic programming and effectiveness [8]. These strategies are supposed to encourage all partners to integrate within a common vision and to develop a creative social capital within the community.

A key premise of the concept of smart specialisation is the application of the „entrepreneurial discovery process" of new domains of specialization [7] and learning processes. This consists in search for new development areas and opportunities, indication of industries that will set the specific countries and regions for success. Such industries include water and sewage management which is an important area falling within the range of smart specialisations at the European, national and regional level. In this scope, it is important to set proper directions of R&D and innovative works. Introduction of innovations in the scope of, inter alia, water recycling and reuse of water in the water industry can bring multiple benefits for sustainable development [13].

An aspect significant for the purpose of increase the water technology development opportunities is cooperation that must be based on network and cluster relations between the various participants of this process at all the different levels. The process of formation and development of smart specialisations depends on participation and involvement of various entities, inter alia, universities, enterprises, public institutions.

In the field of water economy for supporting innovation, development of smart specializations, innovative partnerships at EU level are visible. The Europe 2020 Flagship Initiative for an Innovation Union proposed the concept of European Innovation Partnerships (EIP). European Innovation Partnerships, aim to speed up innovations that contribute to solving societal challenges, enhance Europe's competitiveness and economic growth. EIPs help to pool expertise and resources by bringing together public and private actors at EU, national and regional level, combining supply- and demand-side measures [5].

One such partnership is the European Innovation Partnership on Water – EIP Water in short. The EIP Water facilitates the development of innovative solutions to address major European and global water challenges. At the same time, the EIP Water supports the creation of market opportunities for these innovations, both inside and outside of Europe. EIP Water focuses furthermore on the following 5 key barriers to innovation in the water sector [5]:

- Access to funding and adequate financial instruments
- Overcoming regulatory barriers
- Promoting best practises in public procurement
- Identifying the role of Public-Public/Public-Private Partnerships
- Promoting testing facilities and dissemination of showcases.

The main objective of EIT Water is to initiate and promote collaborative processes for change and innovation in the water sector across the public and private sector, non-governmental organisations and the general public.

Eight priority areas have been chosen for the EIP Water within its Strategic Implementation Plan (Table 1) [5].

Including five thematic priorities and three cutting priorities. They centre on challenges and opportunities in the water sector, and on innovation driven actions that will deliver the highest impact. Smart technology has been defined as an enabling factor for all priorities [5].

Table 1. Eight priority areas have been chosen for the EIP Water

Five thematic priorities	1. Water reuse and recycling 2. Water and wastewater treatment, including recovery of resources 3. Water-energy nexus 4. Flood and drought risk management 5. Ecosystem services
Three cutting priorities	6. Water governance 7. Decision support systems and monitoring 8. Financing for innovation

3 The role of smart specialization in innovative development of a water and wastewater economy in Poland and the Silesian region

Analysis of strategic documents at the national level shows that one of the key technological areas of great significance for the development of Poland is water and sewage management. It is an important research and development area falling within the National Smart Specialisations (NSS) in Poland.

The national smart specialisations are R&D and innovative areas with the highest development potential for the Polish economy and society [14]. NSS consists in focusing investments in the areas ensuring increase of the added value and competitiveness of economy on foreign markets which is to contribute to, inter alia, creation of innovative solutions for socioeconomic development as well as support of transformation towards an economy effectively using resources, including natural raw materials [14].

Water and sewage economy, mostly due to such challenges as limited global availability of natural resources, entails the need of application of more effective and sustainable methods of such management. New innovative solutions in water and sewage economy are necessary to, inter alia, guarantee continuity of supply, effective process coordination, water quality improvement, cost reduction. Use of the cutting-edge innovative solutions has become almost inevitable and is an additional challenge for the companies operating in this industry [15-17].

Water and wastewater economy forms part of one of 17 National Intelligent Specializations, i.e. KIS 9 in the thematic area: Natural Resources and Waste Management. Within the KIS 9. „Innovative solutions and technologies in water and wastewater management", the following are distinguished [14]:

1. Improvement of quality of water for consumption and economic purposes.
2. Increase of water resources for consumption and economic purposes.
3. Improvement of surface and ground water quality.
4. Sewage treatment.
5. Recovery of water and other raw materials from the sewage.
6. Use and recovery of energy in the water and sewage management.

Implementation of National Smart Specialization will take place both through the implementation of national programmes (e.g. NCBiR, PARP projects) and with the use of EU funds under the operational programmes [18]. The system for monitoring the National Smart Specialization (NSS) consist of the following elements [18]:

a) IT platform, which will be basis for the monitoring system
b) Steering Committee (SC)
c) Consultative Group (CG)
d) Economic Observatory (EO)
e) Working Groups for national smart specializations (WG)
f) Regional Forum for Smart Specialization (RFSS).

As part of the NSS monitoring system, inter alia a working group was set up in the area of water and wastewater economy. This group is called KIS-12 "Innovative solutions and technologies in water and wastewater management" [19]. It includes representatives of universities, research institutes, enterprises and business support institutions. The Working Group KIS-12 operates on the wide range of topics of quality improvement and increase of water resources, wastewater treatment, water recovery and other raw materials from wastewater, as well as the use and recovery of energy in water and wastewater management [20]. As part of the activities of the KIS-12 group, the issues were divided into three areas: water, sewage and sediments, and municipal engineering.

The essence of the „entrepreneurial discovery process" is to support bottom-up activities and initiatives that will lead to smart growth and optimal utilization of resources for the development of smart specializations in a given area [18]. Activities implemented at the regional level are of great importance. Water and wastewater economy not only forms a part in the National Smart Specializations, but also often in Regional Smart Specialization. An example of such a region is the Silesian Region. The Silesian Region has the most industrial economy in Poland [21]. Against the state, the Silesian economy has high engagement in research and development (R&D) activities. The water and sewage industry falls especially within the following Regional Smart Specialisation of the Silesian Voivodship: Green economy. Green economy is a priority area for the Silesian Voivodship due to the highest potential for use in the economy, covering measures supporting economic growth and development, concurrently ensuring availability of natural capital and ecosystem services (green growth), environmental, economic and social policy as well as innovations ensuring effective use of resources in the production and consumption process for the society (green economy) [22]. According to the Report "Global Green New Deal. Policy Brief", one of the key areas of green economy is sustainable use of lands, waters and forests. Within the technology groups resulting from the conducted process of entrepreneurial discovery regarding regional smart specialisation "Green Economy", technology groups and subgroups related to the water and sewage economy have been distinguished (Table 2).

Table 2. Selected technology groups resulting from the entrepreneurial discovery process related to water and wastewater economy.

Technology groups	Technology subgroups/ technologies
Biotechnologies for environmental protection	Sewage treatment
Civil engineering technologies	Information materials for the water and sewage industry
Water processing (treatment and separation) technologies, collection and treatment of water	Sewage treatment Water treatment Water and sewage transport systems Waste management
Environmental technologies in various industries	Production and supply with water

The following types of undertakings are distinguished for the purpose of development of this specialisation [22]:

- Innovative projects, including investments in infrastructure conducted by entities and consortia in the technological areas of the specialisation,
- Support of enterprises, especially the SME sector,
- Support of development of participation of the SME sector companies functioning on the green economy market in the regional and supra-regional cooperation networks,
- Support of intentionally recognised scientific and research competence centres based on co-creation and sharing of the research infrastructure,
- Support of functional and operating competence centres promoting specialisation development,
- Support of scientist - entrepreneur contacts, exchange of experiences among entities, search for and improvement of talents for the benefit of specialisation development,
- Activisation of attitudes and prosumer groups in the green economy.

A particular form of involvement of companies in the „entrepreneurial discovery process" is the activity of clusters, bringing together entrepreneurs and representatives of business environment institutions. Cooperation between various partners based on network and cluster relationships is very important. In this area, the Silesian Water Cluster operates in the Śląskie Voivodeship. The Silesian Water Cluster is a joint initiative of water and sewage management enterprises, local authorities and universities having acommon goal of rational water management [21]. The main goal of the Cluster's activity is mutual co-operation aiming to protect, improve the quality and rationally use the available water recourses of the Silesian Province as well as to co-ordinate activities which promote the reliability and safety of water supply systems' functioning and the implementation of sustainable waste water management ensuring protection of water and land ecosystems [23]. The purpose of the Silesian Water Cluster is to create conditions favouring conduction of water-sewage policy according to the regional Development Strategy „Śląskie 2020" [23]. The technological specialization of the entities belonging to the Silesian Water Cluster focuses on implementing innovative solutions in environmental protection, especially in the scope of water purification and waste water treatment, activated carbon regeneration techniques, waste management of sewage and post-coagulation sediments and using renewable energy sources by water and sewage management companies [21].

Scientific and development institutions in the region have a very important role in the scope of environmental technologies. An example of such an organisation is the Institute for Industrialised Area Ecology in Katowice with its Environmental Technology Verification Department [24]. This Department received accreditation for verification of innovative environmental technologies in the water and sewage economy area according to standard PN-ISO 14034:2018-07. Its introduction should contribute to improvement of reliability of information regarding effects of operation of innovative environmental technologies which may facilitate introduction of new solutions both on the national and foreign markets [24].

4 Conclusions

The concept of smart specialization plays an increasingly important role in creating and developing innovations in water and wastewater economy. It is also an important mechanism for the development of innovative business entities. Smart specialisation is of crucial importance to effectiveness of research into and investments in innovation.

Supporting the development of innovative solutions in the field of water-related challenges can bring many economic, social and environmental benefits. The presented analysis indicates the great importance of technologies related to water and wastewater economy, which are an important direction of research and development in the „entrepreneurial discovery process" at the European, national and regional level (an example of the Silesian Region). There is a need to identify factors required to implement the process and to create an effective innovation system that fosters diffusion of innovation. Decisions on smart specializations are the result of in-depth analyzes of endogenous economic advantages and cooperation with socio-economic partners. To a large extent, smart specialisation development possibilities result from the use of partners' research and development capacity. As the results of the analysis show, the water supply industry is in Poland a part of both national smart specializations and regional smart specializations in the Silesian Region.

There is, therefore, a need to seek, establish and develop a strategic partnership between the EU countries and regions. To achieve success in the field of smart growth strategy, it is important to convince all stakeholders that this is "a collective social undertaking" based on the assumption that by working together they can accomplish more than by working separately [7]. Such cooperation is based on trust built on partners reliability as well as observance and pursue of moral values [25]. In order to take full advantage of the opportunities related to water innovation in all related sectors, cooperation at regional, national, supranational and EU level is necessary.

References

1. I. Kłosok-Bazan, *Wdrażanie innowacji w gospodarce wodociągowej*, (Wydawnictwo Instytutu Badań Systemowych Polskiej Akademii Nauk, Katowice-Warszawa, 2009)
2. I. Kłosok-Bazan, B. Gajdzik, J. Machnik-Słomka, W. Ocieczek, Metalurgija **54,** 2 (2015)
3. I. Kłosok-Bazan, Metalurgija **55**, (2016)
4. I. Zimoch, Ochr. Sr. **29**, 4 (2007)
   http://www.os.not.pl/docs/czasopismo/2007/Zimoch_4-2007.pdf
5. EIP Water, https://www.eip-water.eu/
6. J. Machnik-Słomka, I. Kłosok-Bazan, *Wdrażanie innowacji w gospodarce wodociągowej*, (Wydawnictwo Instytutu Badań Systemowych Polskiej Akademii Nauk, Katowice-Warszawa, 2009)
7. *Przewodnik Strategii Badań i Innowacji na rzecz inteligentnej specjalizacji (RIS 3)*, (Luksemburg: Urząd Publikacji Unii Europejskiej, 2012)
8. http://ec.europa.eu/regional_policy/sources/docgener/informat/2014/smart_specialisation_pl, European Commission
9. P. David, D.P. Foray, B. Hall, *Knowledge Economists Policy Brief*, 9 (2007)
10. The Role of Community Research Policy in the Knowledge-based Economy, Expert Group Report, European Commission, Belgium (2009)
11. EUROPA 2020. KOM 2010 (2020), European Commission (2010)
12. D. Foray, J. Goddard, X.G. Beldarrain et all., Guide to Research and Innovation Strategies for Smart Specialisations (RIS 3), European Commission (2012)
13. I. Zimoch, J. Szymik-Gralewska, Ecol. Chem. Eng. S **23**, 3 (2016)
14. Krajowe Inteligentne Specjalizacje, Ministerstwo Rozwoju (2017)
15. I. Zimoch, E. Szymura, K. Moraczewska-Majkut, Desalin. Water Treat. **57**, 3 (2016)
16. I .Zimoch, E. Łobos, Desalin. Water Treat. **52**, 19/21 (2014)
    https://doi.org/10.1080/19443994.2014.884684
17. M. Bożym, I. Kłosok-Bazan, M. Wzorek, Pol. J. Environ. Stud. **27**, 3 (2018)
18. National Smart Specialisation in Poland (2014)
19. http://www.smart.gov.pl/
20. K. Barbusiński, M. Żubrowska-Sudoł, Gaz Woda, 3 (2017)
21. Clusters in the Śląskie Voivodeship, PARP, Warszawa (2011)
22. Model wdrożeniowy Regionalnej Strategii Innowacji Województwa Śląskiego na lata 2013-2020 wersja 2.0, Katowice (2018)
23. Silesian Water Cluster, https://www.clustercollaboration.eu/cluster-organisations/silesian-water-cluster
24. Przewodnik dla wnioskodawców pilotażowego programu Weryfikacji Technologii Środowiskowych (ETV) Unii Europejskiej. AdvanceETV, Katowice (2012)
25. J. Stachowicz, A. Stachowicz-Stanusch, Org. Zarz. **4** (2011)

Disinfection by-products monitoring in water of selected outdoor swimming pools in Opole Voivodship

Iwona Klosok-Bazan[1*], *Marta* Bożym[2], *and Małgorzata* Wzorek[3]

[1] Opole University of Technology, Department of Thermal Engineering and Industrial Facilities, Mikolajczyka 5 St., Opole, Poland
[2] Opole University of Technology, Department of Environmental Engineering, Mikolajczyka 5 St., 45-271 Opole, Poland
[3] Opole University of Technology, Department of Process Engineering, Mikolajczyka 5 St., 45-271 Opole, Poland

Abstract. The aim of this study was to determine the level of disinfection by-products in selected outdoor swimming pool in the Opole Voivodship. The authors paid special attention to the determination of the concentration of trihalomethanes (THMs), which are formed during disinfection in water. Five outdoor swimming pools were selected in five different cities located in Opole Region. The level of trihalomethane (THM) concentrations in the analyzed waters was found to be in the wide range from 15 to more than 550 μg/L. The dominant compound of THMs was chloroform because of chlorine application in the disinfection method.

1 Introduction

Disinfection forms a mandatory process for the public indoor and outdoor swimming pools. Chlorine-based disinfectants are included among the most frequently applied disinfectants and oxidizers for swimming pool treatment.

In contact with swimming pool water, these substances produce a variety of disinfection byproducts (DBPs), such as trihalomethanes (THMs), whose concentration needs to be regulated in swimming pools water. The amount of DBPs that are formed in pool water is influenced by several factors, including free residual chlorine, number of swimmers, water temperature, total organic carbon (TOC), water resource, and pH [1-8]. World Health Organization (WHO) data point out that the toxicity of those hazardous substances introduced by the inhalation route is much higher than by ingestion or through dermal contact. Therefore WHO recommend a maximum concentration of 100 mg/L of THMs for all types of swimming pool waters. Additionally, the International Agency for Research on Cancer (IARC) classified some THMs (chloroform and bromodichloromethane) as potentially carcinogenic to humans (group 2B) [8].

In Poland standards of swimming pool water quality are regulated by the national law, i.e. by the Regulation of Minister of Health on the requirements to be met by water in swimming pools [9]. On the list of the many physico-chemical parameters of water, the parameters relating to DBPs are included as well.

Selected parameters associated with the formation DBPs in swimming pool water are summarized in Table 1. In accordance with WHO recommendations, the maximum level of THMs should not exceed 100 μg/L, wherein chloroform concentration cannot be higher than 0.03 mg/L for all pool types, except for water in the pool basin using by infants. In such facilities maximal chloroform concentration must be lower than 0.02 mg/L.

A variety of studies have been disseminated on the presence of DBPs in swimming pools. Some reports [10-14] have served to offer an account of the total of THMs, or identified only chloroform [15-18]. There are also publications which include measurements that have been carried out with regard to pool water on a reduced scale or with samples generated in laboratory-scale simulations [17-21]. Unfortunately little attention has been paid to investigating the seasonal occurrence of DBPs in outdoor swimming pool waters.

Table 1. Selected physico-chemical requirements for swimming pool water according to Regulation of Polish Minister of Health [9].

Parameters	Unit	Water introduced into the pool basin in the circulation system	Water in the pool basin	Water in the pool basin equipped with aerosol devices	Water in the pool basin using by infants
Cl (bound)	mg/L	0.2	0.3	0.3	0.3
Chloroform	mg/L	0.03	0.03	0.03	0.02
ΣTHM	μg/L	100	100	100	100

* Corresponding author: i.klosok-bazan@po.opole.pl

2 Materials and methods

Five outdoors swimming pools located in various cities located in Opole Voivodship were selected for the study. The selection of the study facilities was affected by the possibility of gaining permission for conducting the tests. A lot of outdoor swimming pools operate in the Opole Voivodship, some of them have been in operation for over 80 years. Many of them do not include a recirculation or filtration systems. The operation of these facilities involves only filling the basin using underground water, manual dosing of chlorine compounds and periodic monitoring of biological water quality. With significant deterioration of water quality, the basin is emptied and after cleaning, filled with fresh spring water. We can assume that this solution leads to formation of DBPs. The technical parameters of five selected research facilities are summarized in Table 2.

Water samples were collected during the summer season at intervals of two weeks; each object was tested four times. All water samples were taken at approximately 30 cm under the water surface. Amber glass bottles (100 mL) with glass screw caps were used for the collection of the treated water. The jars were completely filled to avoid the evaporation of volatile compounds. Samples were not stabilized by addition of any reagents. Samples were stored at <4°C during transport to the laboratory. Immediately after sample collection (1-2h), THMs were extracted using liquid-liquid extraction with penthane (10:1, v:v) at room temperature. Samples were shaken manually (1 min) and extracts were stored in amber glass vials at <4°C until analysis was performed not later than 14 days after the date of sampling. Sampling, extraction and analysis procedures were performed according to the PN-EN ISO 10301:2002 standard. Simple standard of THM (TCM, BDCM, CDBM and TBM) - concentration of 2,000 mg/dm^3 in methanol (Restek) was used.

With individual standards a mixture of THMs and methanol (Merck, GC - grade) was performed.

An intermediate standard stock solution of THMs 20 μg/dm^3 was obtained by diluting the THM standard mixture methanol and was stored at -18°C. Calibration standards were prepared from standard stock solution in penthane (Merck, UniSolv(R)). Sufficient linearity was obtained for all THMs with the correlation coefficients r>0.995.

Separation and identification of THMs was carried out with an Agilent Technologies 7890B gas chromatograph equipped with a micro-electron capture detector (μECD) and a capillary column (Ultra Inert DB-5MS, 30 m×250 mm×0.25 mm). Helium was used as the carrier gas with a flow rate of 1 mL/min, and nitrogen was used as makeup gas with the flow rate of 30 mL/min. The oven was held at 35°C for 20 min, then it was ramped at 10°C/min to 150°C, held for 3.5 min at 150°C. The injector and detector temperatures were 200°C and 300°C, respectively. The injection was operated in a splitless mode.

3 Results and discussion

The concentration of trihalomethanes in water was characterized by a considerable variability in the examined pools. A wide range of concentrations expressed in terms of the total of these compounds was determined, i.e. 15-551 μg/dm^3 (Table 3). The lowest concentration ∑ THMs was found in the samples derived from the pool no. 3, as it was equal to 116 mg/L. Among the facilities applied in this analysis, the swimming pool no. 3 forms the most modern of them all, as it underwent a thorough upgrade in 2006. In this facility, water filtration is used, as it probably affects the limitation of the number of precursors, and thus the formation of THMs (Table 3).

The highest concentration in terms of the total of trihalomethanes was recorded in the water of facility no. 2, as it was equal to 338 μg/L, and the concentration of the total THMs for pool no. 1 was equal to 158 mg/L and

Table 2. Selected technical and technological parameters of outdoor swimming pools.

Technical data of investigated facilities	**Research objects**				
	No. 1	**No. 2**	**No. 3**	**No. 4**	**No. 5**
Volume of the swimming pool, m^3	ab. 1,200	ab. 1,200	ab. 1,200	ab. 1,200	ab. 1,200
Basin dimensions (length *x* width), m	50 x 22	50 x 20	50 x 20	50 x 20	50 x 20
Disinfection method	NaClO	NaClO	NaClO	NaClO	NaClO
Recirculation and water treatment system	none/complet e replacement of water	none/complete replacement of water	filtration and recirculation	none/complete replacement of water	none/complete replacement of water
Other data	built in the 1930s	no data	built in 1950s after a complete refurbishment in 2006	built in the 1970s	built in the 1970s

in the facility no. 4, it was 128 µg/L, compare to 274 µg/L in the facility no. 5. It is not possible to determine a reason to explain the differences in the concentrations of trihalomethanes in the individual pools. We can assume that the content of these compounds is affected by the number of people who use the pools, but this statement could not be confirmed because there is a lack of data regarding the number of swimmers in the examined facilities in the summer. Hence, we can only assume that the level of THMs increases with time, starting from filling the pools and throughout the summer season, especially for facilities that do not include a water treatment system. However, no such relationship was observed.

As we know, the concentration of THMs in water is affected by many factors, not just the content of precursors and the degree of its chlorination, but also water pH, insolation and air temperature [1-3,12,13]. Thus, we can assume that the reason for the high variability of THMs is associated with the difference between the quality of water that is applied to fill the pools. In each of the examined facilities, water came from a different source.

Among the investigated THM_S, chloroform (TCM) was responsible for the highest ratio, as in the pool no. 1 the chloroform concentration was equal to 58 µg/L. In the pool no. 4 the level of chloroform was at a level of 93 µg/L. Pool no. 5 was characterized by the highest levels of the examined parameters, as the concentration of TCM was equal to 223 µg/L in it.

In the case of pool no. 2, a high concentration of bromoform was also recorded, at a level that was similar to the concentration of chloroform. This results could be due to the higher concentration of bromides in the water supplying the pool basin.

The total organic carbon (TOC) determined in the waters of the examined facilities did not differ significantly, i.e. 2.1–3.8 µg/dm^3. The concentration of TOC in the examined waters was found not to affect the concentration of trihalomethanes.

For the purposes of gaining comparative results, the data regarding outdoor swimming pool was compiled with data for indoor pools located in the Opole Voivodship, with a note that the measurements were carried out at the same time. Table 4 shows the mean results for chloroform and total THMs coupled with the data regarding the range of minimum and maximum concentrations for five indoor and outdoor pools. As we can see, the values recorded in the indoor facilities were considerably lower. Indoor pools are modern facilities that contain water treatment systems, which affects THMs levels in waters.

Table 3. The results of the analysis of THMs in 5 selected outdoor swimming pools waters in Opole Voivodship.

Parameters	**Research object**				
	1	**2**	**3**	**4**	**5**
TCM, µg/dm^3	136±78 (57-214)	153±75 (41-194)	58±21 (37-78)	92±30 (58-116)	223±153 (15-372)
BDCM, µg/dm^3	7±1 (5-8)	5±4 (3-10)	1±0.1 (<0.1-1)	8±2 (5-9)	10±7 (<0.1-16)
DBCM, µg/dm^3	1±0.1 (<0.1-1)	1±1 (<0.1-2)	1±0.1 (<0,1-1)	<0,1	1±0,1 (<0.1-2)
TBM, µg/dm^3	14±9 (4-20)	179±126 (54-351)	57±42 (21-114)	28±11 (19-40)	40±62 (<0.1-130)
∑THM, µg/dm^3	158±79 (83-241)	338±148 (233-551)	116±60 (59-190)	128±42 (82-164)	274±210 (15-518)
TOC, µg/dm^3	3.0±0.3 (2.7-3.2)	2.1±1.5 (1.1-3.2)	3.5±0.4 (3.2-3.8)	3.8±0.8 (3.2-4.4)	3.7±0.7 (3.2-4.3)

Table 4. Results of water samples from the selected outdoor and indoor pools in Opole Voivodship.

Parameter	**Outdoor pools**		**Indoor pools [10]**	
	Concentration, µg/dm^3	**The proportion of samples exceeding the norm value ***	**Concentration, µg/dm^3**	**The proportion of samples exceeding the norm value ***
Chloroform (TCM)	132±63 (15-372)	94%	61±47 (26-208.2)	76%
∑THM	203±98 (15-551)	78%	122±74 (28-279)	57%

*- 100 mgTHM/dm^3 and 30 mgTCM/dm^3 according to [9]

4 Conclusions

By-products of water treatment are formed in the waters of outdoor pools in Opole Voivodship, mostly ones containing chloride derivatives of methane and ones taking the form of chloroform. The identified contamination levels in each of the analyzed outdoor pools exceed the values specified in the regulation of the Minister of Health stipulating the requirements regarding swimming pool water. It is significant that the tests were carried out in the summer season preceding the implementation of the above regulation as valuable source of information was gained for the operators of the examined facilities. All examined facilities had to implement corrective action plans, the outcome of which involved the decrease of the total concentration of THMs to less than 100 μg/dm^3. The results obtained and reported in this paper do not deviate substantially from the results of studies reported in scientific journals. Thus, we can state that the maintenance of a low concentration level of water disinfection by-products is a difficult task – especially in facilities where the only method applied to ensure microbiological safety of water involves batching chloride compounds. The type of pool appears to have an impact on the levels of recorded DBPs. The measured concentrations of THMs are greater in outdoor pools.

References

1. S. Simard, R. Tardif, M.J. Rodriguez, Water Res. **47** (2013)
2. I. Klosok-Bazan, Ochr. Sr. **36**, 2 (2014)
3. I. Zimoch, E. Szymura, K. Moraczewska-Majkut Desalin. Water Treat. **57,** 3 (2016)
4. E. Righi, G. Fantuzzi, G. Predieri, G. Aggazzotti Microchem. J. **113**, 23 (2014)
5. I. Zimoch, E. Łobos, Desali. Water Treat. **52**, (2014) https://doi.org/10.1080/19443994.2014.884684
6. I. Zimoch, Ochr. Sr. **29**, 4 (2007) http://www.os.not.pl/docs/czasopismo/2007/Zimoch_4-2007.pdf
7. C.M. Villanueva, L. Font-Ribera Annali dell'Istituto Superiore di *Sanità* **48**, 4 (2002)
8. *International Programme on Chemical Safety* IPCS, "Disinfectants and disinfectant by-products. International programme on chemical safety," Environmental Health Criteria Geneva, World Health Organization, Geneva, Switzerland, 2000.
9. Regulation of Minister of Health from 9 November 2015 on the requirements to be met by water in swimming pools [In Polish]
10. M. Bożym, I. Kłosok-Bazan, M. Wzorek, Pol. J. Environ. Stud. **27**, 3 (2018)
11. J. Boguniewicz-Zablocka, I. Klosok-Bazan, V. Naddeo, Environ. Sci. Pollut. Res. (2017). (doi.org/10.1007/s11356-017-0608-8)
12. H. Kim, J. Shim, S. Lee, Chemosphere **46**, 1 (2002)
13. S. Simard S.R. Tardif, M.J. Rodriguez, Water Res. **47**, 5 (2013)
14. I. Zimoch, Ochr. Sr. **33**, 3, pp. 35-42 (2011) http://www.os.not.pl/docs/czasopismo/2011/3-2011/Zimoch_3-2011.pdf
15. G. Aggazzotti, G. Fantuzzi, P. Tartoni, G. Predieri, Arch. Environ. Health, **45**, 3(1990).
16. C. Catto, G. Charest-Tardif, M. Rodriguez, R. Tardif Environ. Pollut. **1**, 2 (2012)
17. M.J. Chen, C.H. Lin, J.M. Duh, W.S. Chou, H.T. Hsu, J. Hazard. Mater. **185**, 2 (2011)
18. B. Lévesque, P. Ayotte, R. Tardif, G. Charest-Tardif, E. Dewailly, D. Prud'Homme, J. Toxicol. Environ. Health A **61**, 4, 225 (2000)
19. M.J. Farré, D. Stalter, J.Y. Tang, J. Molendijk, B.I. Escher, **59**, 172 (2014)
20. T.L. Teo, H.M. Coleman, S.J. Khan, Environ. Int. **76** (2015)
21. K. Mizerna, A. Król, A. Mróz, E3S Web of Conferences, **19** (2017)

Testing the efficiency of water aeration with the use of ATOL-OXY type air applier

Anna Głowacka[1,*], and *Bartosz Bogusławski*[1]

[1]Department of Sanitary Engineering, West Pomeranian University of Technology in Szczecin, al. Piastów 50, 70-311 Szczecin, Poland

Abstract. The purpose of this study was to determine the effectiveness of aeration with the use of an applier (a device for aeration of sludge protected by the patent application no P-416711) by determination of characteristics of its operation depending on the stream of water flowing through the device, the height of the outlet over the water table and multiplication factors of measuring water pumped through the applier. On the basis of laboratory tests/measurements, a mathematical model of the increased concentration of oxygen dissolved in water has been developed.
As a result of the research, a significant increase in oxygen concentration has been achieved only after the first water pumping through the applier. The determined mathematical model can be used to calculate predicted increases of oxygen concentration dissolved in water in variants other than those investigated.

1 Introduction

The main problem in the design of sanitary sewage systems is the decreasing water consumption for social welfare purposes, which in consequence reduces the content of the input oxygen dissolved in wastewater while increasing the concentration of pollutants. This problem is particularly evident in rural and peripheral systems, where small wastewater flows in the ducts increase the stopping time in pumping stations and discharge pipelines. Too low a flow velocity (less than 0.8 m / s - considered self-cleaning velocity of the ducts) causes sedimentation, deposition, and decay of organic deposits at the bottom of the pipelines [1-8].

Introduction of general water consumption metering under the Act of 7 June 2001r. about collective water supply and collective waste water discharge (Journal of Laws of 2001, No. 72, item 747) [9] caused the population's water consumption to decrease markedly. In 1990, household water consumption was 192,300 m^3, 136,000 m^3 in 2000, and 119,600 m^3 in 2014 [10]. Reduced water consumption by the residents results in reduced effluents with constant production of pollutants resulting from human metabolism, and consequently lower flow rates and velocities in sewage channels [11]. Smaller amounts of sewage in the sewage system (especially in very dispersed systems) result in long flow times, and in particular long drain times at local pumping stations. Recommended in the German standard [12], the time of sewage retention in the pumping station is 4 hours. After this time, the sewage rots, generating odors and is dangerous to health due to the hydrogen sulphide evolution [13-18].

An attempt to counteract the formation of oxygen deficits in sewage systems is the use of an air applicator type ATOL-OXY. The tests were aimed at checking the aeration efficiency of the device.

2 Materials and methods

ATOL-OXY air applier is a device designed for water and sewage aerification. The operation principle of the device is to disperse a stream of liquid in the form of a cone in the air, which causes repeated air increase in contact with air. For lower flow values and higher unit heights above the water table, the cone surface is interrupted to form drops.

The research on content of oxygen soluble in water was conducted between July 21st and September 28th 2015. A total of 96 measurements of the oxygen concentration dissolved in water were performed.

The concentration of dissolved oxygen was performed According to Water Quality PN-EN 25813. Determination of dissolved oxygen. The iodometric metod [19] was performed in a laboratory of Sanitary Engineering Department of the Water Technology, Wastewater and Waste Department.

2.1. Description of the measuring station operation

The measuring station is shown on Fig. 1. In the A tank with the capacity of 1000 dm^3 a pump was placed (1) resposible for the production of kinetic energy and circulation of measured water. Water was transported by a DN 50 steel pipe to a ball valve (2), with which the flow was stifled. By controlling the indicators on the water meter (3) the desired flow rate was set through the valve. The section from the valve to the water meter (2-3) was made of PN10 pipe with the diameter of

* Corresponding author: anna.glowacka@zut.edu.pl

40x2.5 mm PE. Further sections from the water meter to the applier were made of PN10 pipe with the diameter of 50x3mm PE. Manometer (4) performed a control function. Next, water flowed through the applier and got into the B tank from which samples were taken in order to determine the amount of dissolved oxygen. Then, water returned to the A tank by a gravity duct.

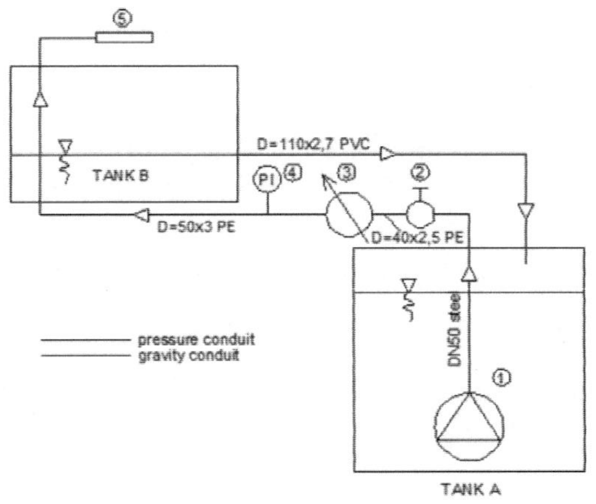

Fig. 1. Scheme station diagram: 1 - Grundfos Unilift AP35 submersible pump; 2 - ball valve Gebo DN 50; 3 - Techmag FMG 300k DN 40 water meter; 4 - manometer (measuring range 0-1 bar); 5 - ATOL-OXY Magnus-Tech air applicator.

The construction of the measuring station and tested device were developed in cooperation with Magnus-Tech from Szczecin and a designer Eng. Wojciech Bogusławski.

The measurement sequence was divided into 3 series. Each series consisted in determining the amount of oxygen dissolved in water after the water passed through the applier for subsequent flow values 1; 1.5; 2; 2.5; 3 dm^3/s at a specified applier's height above the water table in the B tank. The first one was held at the height of 0,3m, and the next at 0.4m and 0.5m. Each series preceded the replacement of water in the A tank for fresh tap water.

The measurement series started with the delivery of 800 dm^3 tap water to the A tank.

Each measurement cycle was started from water deoxidization in the A tank with sodium sulfate (Na_2SO_3) and then a sample of "0" was taken to determine the initial oxygen concentration. All samples were taken into conical flasks with the capacity of 100ml in 2 repetitions. For a given flow value, the entire volume of water in the A tank was passed through the applier 5 times. Each time water passed through to the B tank a sample was taken in order to determine the concentration of dissolved oxygen. As the water was constantly circulating in a closed circulation, each flow was characterized by different time of overflow of the entire water volume. In a single measurement cycle, one flow value was tested.

Water overflow time from the A tank for all flow values is shown in Table 1.

To make sure that all water in the B tank is oxygenated, samples were taken at the end of each time water passed through the applier. During measurement the overpressure value on the manometer and the test water temperature was recorded, which averaged 21.16 ± 1.09°C.

Table 1. Water overflow time for the tested flow values.

Flow	Time one ribbon	
[dm^3/s]	[s]	[min]
1	800	13:20
1.5	533	8:54
2	400	6:40
2.5	320	5:20
3	267	4:30

4 Results and discussion

Basing on the measurement data, the operating characteristics of the device were determined in relation to the height above the water table, the flow rate and the multiplication of the water flow through the applier. In addition, a model of the increase of dissolved oxygen concentration after one-time water overflow was calculated as a function of two variables - height and flow.

Analyzing the increase in oxygen concentration dissolved in water after one-time overflow through the applier depending on the flow rate for the three tested heights: 0.3; 0.4 and 0.5 m of the applier above the water table, an upward trend in oxygen concentration was proven which was dependent on the flow and height of the applier over the water table (Fig. 2). It was observed that the highest increase in water soluble oxygen was at the height of 0.5 m and amounted to 5.05 mgO_2/dm^3 (flow 1 dm^3/s), and the lowest increase was at the height of 0.3 m - 2.75 mgO_2/dm^3 (flow 1 dm^3/s). In subsequent calculations of the mathematical model of operation after a measurement error was defined, the measurement result for the flow of 2.5 dm^3/s at H = 0.4 m was rejected. As a result, regression equations were estimated and an approximate linear increase in dissolved oxygen concentration was observed at one time, depending on the flow rate (Fig. 3).

In further research, basing on the laboratory experiment, a mathematical model was developed. It described the concentration dependence of dissolved oxygen in water from the height and flow of water. To define a complete model of oxygen concentration increase, a linear regression was used as a function of two variables and the below function pattern was obtained:

$$\Delta C = f(H, Q) = 8{,}01H + 0{,}842Q - 0{,}172 \qquad (1)$$

Description of equation:

ΔC - increase in oxygen concentration, mg/dm^3

H - outlet height of the applier over the water table, m

Q - water flow through the applier, dm^3/s

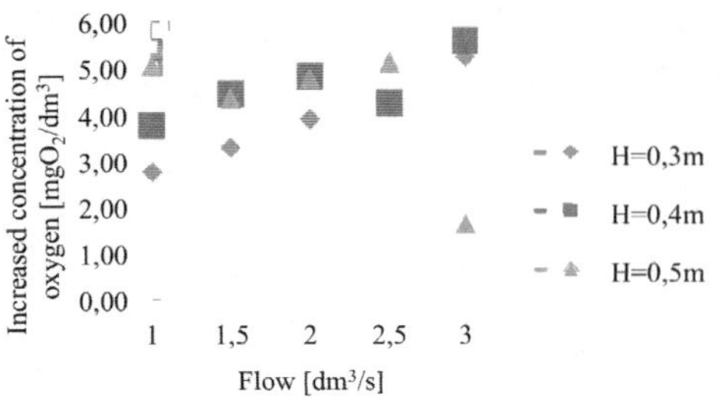

Fig. 2. Increased concentration of oxygen dissolved in water depending on flow rate and applicator height above the water table.

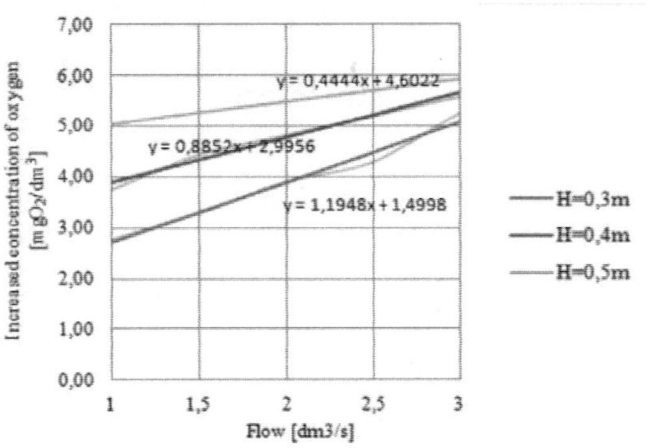

Fig. 3. Trend lines for the increase of oxygen concentration depending on the flow and the height of the applicator, once the water has been overpressed by the device.

Based on the mathematical model, the relation shown in Fig. 4 was developed.

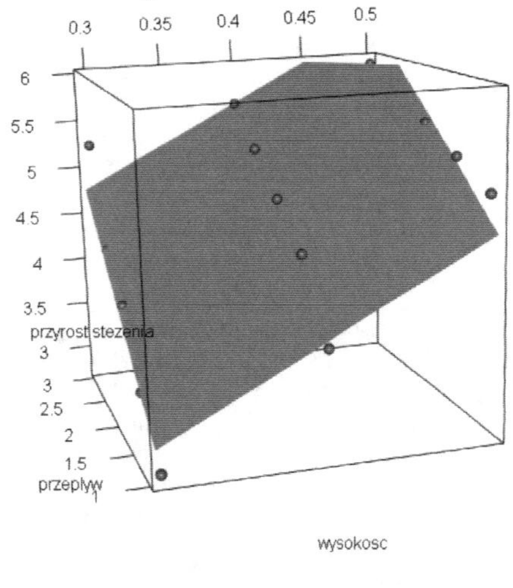

Fig. 4. Increase in oxygen concentration depending on altitude and flow.

According to this dependence, the increase in dissolved oxygen concentration increases linearly as a function of the height of the applier over the water table and the water flow through the applier. This model helped to generalize the three linear functions shown in Figure 3 into one plane of the increase in oxygen concentration. The model also allows to determine the increase in oxygen concentration in different height and flow variants and thus optimize the performance of the device and its full potential.

5 Conclusions

1. After the first injection of water through the applier, a significant increase in oxygen concentration was achieved. The increase in oxygen concentration after single extrusion ranged from 1.63 mg / dm^3 (0.5 m height, 3 dm^3/s flow) to 5.59 mg/dm^3 (height 0.4 m, 3dm^3/s flow).
2. After five times of water overflows, oxygen saturation exceeded 95% in twelve out of sixteen measurements.
3. The mathematical model allows calculation of predicted increases in dissolved oxygen concentration in water in variants other than those tested. In addition, the mathematical model can be used to optimize the operation of the pump-applier system, taking into account the conditions of pumping stations or expansion wells.
4. The use of the applier in existing sewage systems would contribute to the improvement of the aerobic condition of sewage. It would also help solve the problem of oxygen deficiency, the formation of toxic hydrogen sulphide and nuisance odors in both existing and planned sewage systems.

References

1. W. Chełmicki, *Woda - Zasoby, degradacja, ochrona*, (PWN, Warszawa, 2002)
2. J.R. Dojlido, *Chemia wód powierzchniowych,* (Ekonomia i Środowisko, Białystok, 1995)
3. Z. Hiedrich, M. Kalenik, J. Podedworna, G. Stańsko, *Sanitacja Wsi*, (Seidel-Przywecki Sp. z o.o., Warszawa, 2001)
4. J. Łomotowski, Wodociągi Kanalizacja **4**, (2011)
5. H. Sawicka - Siarkiewicz, P. Błaszczyk., N. Gmitrzuk, T. Witarzewska, *Leksykon Ochrony Wód*, (Instytut Ochrony Środowiska - Państwowy Instytut Badawczy, Warszawa, 2011)
6. T. Żaba, A. Królikowski, J. Królikowska, J. E. Eng. **24**, (2011)
7. J. Żelazo, Z. Popek, *Podstawy renaturyzacji rzek*, (SGGW, Warszawa, 2002)
8. http://www.fao.org/docrep/x5744e/x5744e0m.htm# 1.2.2
9. The Act of June 7, 2001 on collective water supply and collective sewage disposal (Dz. U. z 2001 r. nr 72 poz. 747)
10. *Ochrona Środowiska 2014.* (Roczniki Statystyczne GUS, Warszawa, 2015)

11. J.L. Huisman, N. Weber, W. Gujer, Water Res. **38**, (2004)
12. German Design Standard DIN EN 1671
13. J. Worona, Ochr. Srod. **5**, (2012)
14. J.H. Carpenter, Limnol. Oceanogr. **11**, 2 (1965)
15. M. Ochowiak, L. Brodniarz-Press, A. Kaczmarek, N. Kujawa, Technol. Prog. Food Process. **1**, (2013)
16. V. Rathinakumar, G. Dhinakaran, R. Suribabu, JESTEC **12**, 3 (2017)
17. M. Emiroglu, A. Baylar, J. Eng. Env. Sci. **27**, (2003)
18. V. Rathinakumar, G. Dhinakaran, R. Suribabu, P. Velmurugan, AJAS **7**, 6 (2014)
19. Polish Design Standard PN-EN 25813:1997-12

Changes of catalytic mass grains morphology as a result of their surface covering with iron compounds

Maciej Dobrzański [1,*], and *Andrzej* Jodłowski[1]

[1]Lodz University of Technology, Institute of Environmental Engineering and Building Installations, 90-924 Lodz, Politechniki Ave 6, Poland

Abstract. There were operational problems of not having the ability to remove manganese compounds in one of the country's groundwater treatment plants, despite the fact that the filters were some years earlier fitted with the Multiman 3M catalytic bed inserts. The paper presents the results of analysis of morphological characteristics of fresh and blocked catalytic grains. An equivalent diameter, circularity, convexity, and compactness were used to describe the morphological characteristics of the catalytic mass grains. The obtained results confirmed the assumption that the grains of the catalytic mass were blocked by iron hydroxide and that the cause of WTP operating problems was the failure of the aerator, which resulted in the penetration of iron hydroxide through the upper sand bed layer into the lower catalytic mass layer.

1 Introduction

Groundwater is one of the main sources of tap water, especially in rural areas. The principal problem of groundwater treatment is usually the need of excessive concentrations of iron and manganese ions removal. Removal of Fe (II) and Mn (II) ions from water is usually achieved by oxidizing them to insoluble forms, which are then removed in the single- or two-stage filtration process through single or multi-layer beds. The iron removal process is relatively simple, usually involving the use of aeration, followed by filtration through a bed of quartz sand to retain the iron hydroxide precipitated from the water. It is much more difficult to remove manganese from water, whose oxidation requires alkalization of water to a pH of about 9.5 [1]. Adjustment of pH value is not necessary in the case of catalytic active masses. Oxidation of manganese in the catalyst bed is effective only after primary removing excess of iron from the water. Otherwise, the iron hydroxide can be precipitated and retained in the catalytic layer, which significantly reduces the manganese removal efficiency [2,3].

One of the WTP having maximum production of 37 m^3/h use underground water and supplies treated water to rural distribution network. The water withdrawn from the deep well by the deep-well pump is pressed into the pressure aerator, from where through a set of three iron and manganese removing pressure filters it reaches the pressure booster system that provides the appropriate pressure in the distribution network. The filters are equipped with quartz sand beds of approx. 1.0 m height to ensure the retaining of precipitated iron hydroxide as a result of the aeration and at the bottom of the bed of the catalyst mass of Multiman 3M, approximately 0.4 m high. The Multiman 3M catalytic mass responsible for the removal of excessive manganese content is a filtration material whose main component is the MnO_2 manganese ore (MnO_2 content of minimum 82%).

After several years of operation, it was found that the concentration of manganese in treated water was more than three times the permissible value. Then the operator attempted to improve the efficiency of the water treatment by increasing the intensity and frequency of the filter backwashing. It was found that there was a failure of the aerator consisting in tearing the air dispersing grate. This was most probably due to the corrosion and dynamic loads of the aerator components. After the repair, further water quality tests were carried out, the results of which indicated an improvement in the removal of iron from water. The concentration of total iron reached the level below the limit value. The efficiency of manganese removal has not improved, however. The concentration of manganese level of treated water was maintained at 200 μg Mn/L, which was four times the acceptable value.

This paper presents the results of evaluation of the morphological characteristics of catalytic grains collected in a small rural WTP treating underground water, where the removal efficiency of manganese and iron was lost.

2 Materials and methods

2.1 Materials

The comparison of the morphological parameters of the fresh catalytic mass of Multiman 3M and the mass from the filters which did not provide the removal of

* Corresponding author: maciej.dobrzanski@p.lodz.pl

manganese from water, was made. According to the product data sheet specifications, the grains of the catalytic mass are characterized by irregular shape, rough surface and sharp edges. The granular mass of Multiman 3M in the form of brown and black granules is typically 1.0 to 3.0 mm or 0.8 to 2.5 mm. The bulk density of the material is about 2.0 Mg/m^3 at a specific density of 4.0-4.2 Mg/m^3.

2.2 Analytical methods

Measurements and analysis of equivalent diameters and morphological parameters of catalytic mass grains were carried out using Malvern Morphologi G3 microscope image analyzer. As the morphological parameters were used equivalent diameter (d_e, mm) - the diameter of the circle with the same surface as the image of the recorded particle, circularity - the ratio of the perimeter circle to the same surface as the particle divided by the perimeter of the actual particle image (assumes values from the nearest 0 to the value 1.0 for the ideal circle), solidity - the actual surface of the particle divided by the surface of the tensile thread that enters the particle (assumes values from 0 to 1.0 - the ideal circle), elongation - takes values from 0 (ideal circle) to 1.0, convexity - the ratio of the circumference of the tensile thread entangling the particle to the circumference of the actual particle (assumes values from 0 to 1.0 - perfectly smooth surface).

Additional research has also been carried out including bed porosity, bulk density and uniformity coefficient. Porosity of the bed was determined using the relationship (1).

$$\varphi = \frac{V_c}{V} = \frac{Q - Q_s}{V \cdot \rho} \cdot 100 \quad [\%] \qquad (1)$$

in which φ means porosity of the bed (%), ρ is specific density of the liquid phase in the bed (g/cm^3), V is bed volume (cm^3), V_c refers to volume of liquid phase in the bed (cm^3), Q is mass of the bed (g), and Q_s is the mass of solid phase in the bed (g).

Bulk density was determined according to the standard PN-EN 1097-3: 2000 [4].

The uniformity coefficient of grain size (U) was calculated taking into account size of the grains together with the smaller ones accounted for 10% of the weight (d_{10}, mm) and the size of grains, which together with the smaller ones account for 60% of the weight (d_{60}, mm).

3 Results and discussion

3.1 Results of morphological analysis

The morphological analysis of fresh catalytic mass grains showed that the material used for the study was dominated by grains with an equivalent diameter (d_e) of 1.0 - 3.0 mm, which confirmed the information provided by the producer. The sample also had grains of lesser diameter equivalent ($d_e < 1.0$ mm), probably due to the fragmentation of the base material. They were, however, a small part of the volume of catalytic mass (about 3%). Based on a statistical analysis of the distribution of values d_e, it was found that the equivalent of the grain is logarithmically normal (Fig. 1a).

In order to determine the distribution parameters, logarithm of the equivalent diameter values were calculated and it was found that the expected value of d_e was of 1.67 mm with the standard deviation σ = 0.28 mm. Based on the results of the analysis of the catalytic mass extracted from the iron and manganese removal filter which was installed in the WTP where the loss of iron and manganese retention capacity has been observed, it was found that the equivalent diameter of the grains of the extracted material was mainly in the range 1.0 – 3.3 mm (Fig. 1b).

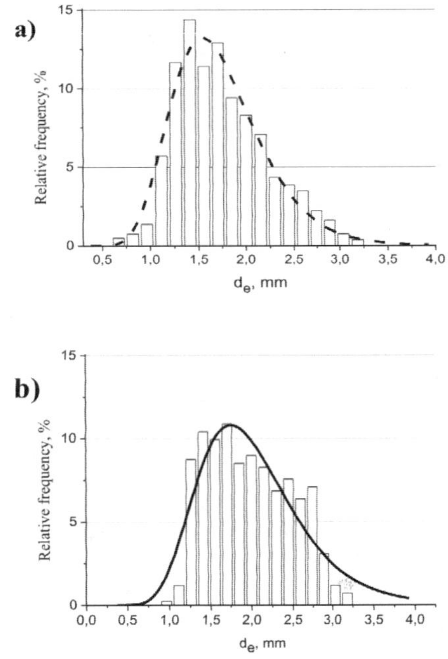

Fig. 1. Distribution of grains equivalent diameter (d_e) of (a) fresh and (b) blocked catalytic mass Multiman 3M.

No grains with a diameter of less than 1.0 mm have been found with such share as in raw material. Probably these small grains were removed from the filter bed as a result of its backwashing. As with the new material, the granulation of the spent bed material was subject to log-normal distribution. The expected d_e value of 1.93 mm was determined and the standard deviation σ was 0.42 mm.

Comparing the results obtained for the blocked catalytic grains bed sample and for the fresh material, the expected value of the distribution of d_e from 1.67 mm to 1.93 mm can be found with the increase in the size of large grains (d_e > 2.0 mm). This confirms the initial assumption that the coating of the catalytic grains with a thin layer of precipitated iron hydroxide should result in an increase in their equivalent diameter.

Based on the cumulative mass distribution of the sample of fresh material, a diameter value of d_{10} of 1.40 mm and d_{60} of 2.22 mm were determined (Fig. 2). By specifying the grain size U for a given sample 1.59 it can be stated that the material to be tested is coarse (equality for $1<U<5$). In the case of blocked mass, the characteristic particle sizes d_{10} and d_{60} and were 1.60 mm and 2.48 mm, respectively, and U-values of 1.55, which also indicates the coarseness of the material being tested.

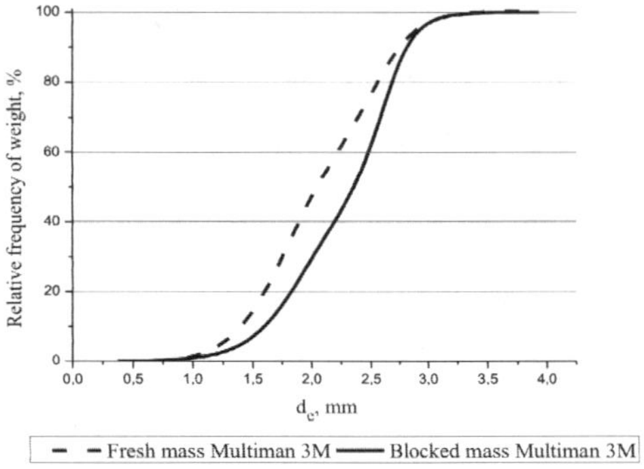

Fig. 2. Cumulative distribution of grains equivalent diameter (d_e) of fresh and blocked catalytic mass Multiman 3M.

The greatest differences between the two samples under the comparison were found based on the results of the analysis of morphological parameters such as circularity, elongation, compactness, and convexity. On the basis of data obtained for both samples it was found, that morphological parameter distributions can mostly be presented using normal distribution. Coating the catalytic grains surface with a layer of iron hydroxide resulted in the shape of the grains being close to spherical. This is illustrated by a change in the expected value of the circularity parameter from 0.85 (fresh catalytic mass) to 0.93 (blocked mass) and the whole distribution of the circularity parameter for this sample shifts to 1.00 (Fig. 3a).

Comparing the values of the convexity parameter (Fig. 3b) determined for the mass of fresh and blocked catalytic mass, the value range was narrowed and the distribution was shifted toward the value of 1.00. In the case of the fresh catalytic mass, the value of the convexity parameter was between 0.91 and 0.97 and of the worn out mass were significantly narrower and ranged from 0.97 to 0.98.

Significant change in morphological characteristics of fresh and blocked mass was noted in the parameter solidity (Fig. 3c). The value of this parameter fitted normal distribution and consisted mainly of between 0.91 and 0.97 in the case of the fresh catalytic mass. In the case of a blocked catalytic mass value of this parameter shifted to 1.00. The sample consisted predominantly of grains having the value of solidity between 0.96 and 1.00.

On the basis of differences in the values of the convexity and solidity parameters, it can be stated that the surface of catalytic grains covered with iron hydroxide had significantly reduced the active area. Grains covered with hydroxide layer had a much more smooth and regular surface. This negatively affects the ability to remove manganese from water.

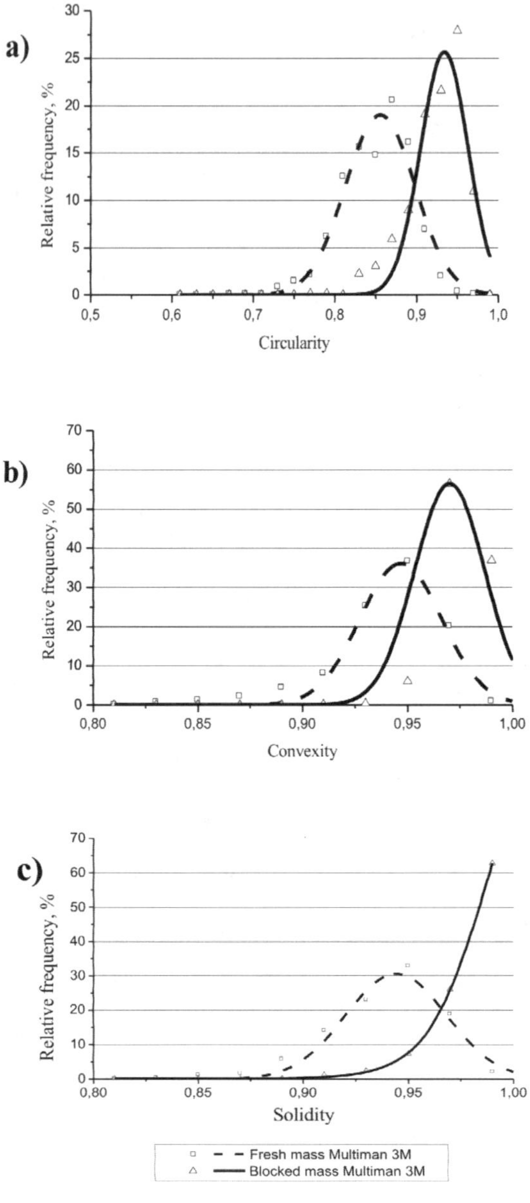

Fig. 3. Distribution of grains parameters circularity (a), convexity (b) and solidity (c) of fresh and blocked catalytic mass Multiman 3M.

List of the fresh (F) and blocked (B) Multiman 3M catalytic mass grains morphological parameters is presented in Table 1.

Photographs taken with the use of the image analyzer enabled the assessment of the grain surface covered by the layer of accumulated iron hydroxide (Fig. 4).

Table 1. Listing of the fresh (*F*) and blocked (*B*) Multiman 3M catalytic mass grains morphological parameters.

Distribution parameter		Expected value	Standard deviation, σ	Percentile	
				90	10
Equivalent diameter, d_e, mm	*F*	1.67	0.28	2.44	1.23
	B	1.93	0.42	2.71	1.33
Circularity	*F*	0.85	0.04	0.90	0.80
	B	0.93	0.03	0.97	0.87
Convexity	*F*	0.94	0.02	0.97	0.91
	B	0.98	0.02	0.98	0.97
Solidity	*F*	0.94	0.02	0.97	0.90
	B	0.98*	0.02*	1.00	0.96
Elongation	*F*	0.25	0.13	0.43	0.09
	B	0.20	0.12	0.37	0.08

* without fitting the distribution

Fig. 4. Pictures of the surface of the blocked catalytic mass with the blocking layer of iron hydroxide.

The deposit cross section shows that it consists of several layers alternately arranged in light and dark material. This can be attributed to the long-term deposition process.

It can be assumed that the light-colored layer is iron hydroxide separated by layers of dark material. Probably it is a dust of catalytic mass from the abrasive grains, which was formed during the intense backwashing that was as attempt of trying to improve the situation.

3.2 Changes in the filtration bed bulk density and porosity as a consequence of the ferric hydroxide layer deposition

Changes in the diameter and morphological parameters of the catalytic mass have resulted in a change in the bulk density. For the fresh catalytic mass it was of 1.72 Mg/m^3. According to the manufacturer the bulk density is about 2.00 Mg/m^3. The mass blocked with iron hydroxide decreased to the value of 1.42 Mg/m^3. Increasing the compactness and circularity of grains and their diameter probably resulted in larger spaces between the grains and thus resulted in a lower bulk density. In order to verify the validity of this thesis the porosity of both masses was measured. In the case of a fresh mass. the porosity amounted to 38%. while in the case of used mass amounted to 49%. The increase in porosity of the bed proves that the proportion of free space in the total volume of the bed increased in the mass blocked with iron hydroxide in comparison with fresh mass.

The close relationship between the porosity of the bed with the size of its grains has been demonstrated by Nowak [5]. who stated that as the active mass grains diameter increases. the porosity of the bed increases. For example. the smallest porosity (41%) was determined for Purolite MZ-10 with a grain size of less than 1 mm. while the largest (52%) for the Defeman catalytic mass. which 90% of grains had a diameter greater than 1.6 mm.

4 Conclusions

The coating of the catalytic grains with the precipitated iron hydroxide eventually led to an increase in their diameter. a decrease in porosity and an active surface. The thickness of deposit layer accumulated on the surface of the catalytic mass grains was estimated at 102 to 284 µm. Over time, the ability to remove manganese from water has been almost completely lost. The greater porosity and diameter of the bed grains of the blocked catalytic mass has probably resulted in smaller flow resistances. Those in turn meant higher filtration velocity. which ultimately led to "break through" the bed and worsened the efficiency of iron removal.

References

1. M.K. Doula, Wat. Res. **40**, 17 (2006)
2. T. Siwiec. Przegl. Nauk. Inż. Kształt. Środ. **53,** (2011)
3. M. Kalenik. D. Morawski, Polska Akademia Nauk. **IV**, 3 (2013)
4. PN-EN 1097-3:2000 Tests of mechanical and physical properties of aggregates - Determination of bulk density and cavity.
5. R. Nowak. J. Ecol. Eng. **17**, 5 (2016)

Differential absorbance as evaluation method of TOX concentrations generated during surface water chlorination

Andrzej Jodłowski [1,*]
[1] Lodz University of Technology, Institute of Environmental Engineering and Building Installations, 90-924 Lodz, Politechniki Ave 6, Poland

Abstract. Chlorination of waters containing natural organic matter (NOM) produces disinfection by-products (DBPs) including harmful THMs and HAAs. A collective parameter and toxicity index of all chlorination products contained in finished water is the total concentration of organic-bound halogen compounds (TOX). Determination of this indicator is carried out by adsorption-pyrolysis method, which is complicated and requires specialized apparatus. The competitive method of TOX concentration determination in water is differential absorbance. Effect of natural organic matter removal from surface waters in coagulation as well in coagulation and adsorption processes on the decrease of TOX concentration measured by differential absorbance was determined.

1 Introduction

The basic purpose of chlorination of water intended for consumption is disinfection, i.e. the destruction of pathogenic organisms. If the water contains organic compounds, chlorine reacts with them, leading to their chlorination and oxidation. Singer and Chang [1] reported that as a result of chlorination of surface waters treated by conventional coagulation, sedimentation and filtration at neutral pH, the THM share of the total number of DBPs formed as TOX after a reaction time of 72 h is only 26%. The ratio of TOX to THM ranges from 3 to 10, depending on the quality of the raw water and the conditions of purification [2].

In order to reduce the potential for DBPs formation, it is usual to use NOM removal prior to water chlorination using coagulation and flocculation followed by sedimentation (or flotation) and filtration. However, it should be keep in mind that the more sensitive to the removal in the coagulation process are high molecular weight NOM components than low molecular weight hydrophilic compounds. Remaining in treated water, they are the precursors of DBPs. The parameter for evaluating the susceptibility of NMO components to coagulation removal is SUVA (Specific UV Absorbance) calculated as UV_{254} (m^{-1}) normalized to DOC (g C/m^3) [3]. If water is characterized by a SUVA value of approximately 2 L/(mg C×m), it means that it contains low molecular weight of non-humic substances and hydrophilic properties. In this situation, powdered activated carbon (PAC) can be used to intensify the removal of NMO-containing substances by coagulation [4].

The method used for the TOX assay presented in Standard Methods [5] (method 5320) includes adsorption, pyrolysis and titration. This method was used for instance by Archer and Singer [6] in the study of the role of SUVA and inhanced coagulation in the removal of TOX precursors. This method is quite cumbersome and requires the involvement of advanced measuring equipment.

Another approach to TOX was proposed by Korshin et al. [7]. Based on the spectrophotometric analysis of DBPs precursors compounds, they suggested the possibility of using the method of estimating the concentration of total halogenated organic compounds (TOX) in the method of differential absorbance. Contrary to the conventional NOM absorption spectra before and after chlorination, UV diffraction spectra have almost always a peak near 272 nm. The TOX method requires a spectrophotometer with a measuring range in UV and by measuring the absorbance at 272 nm before and after the chlorination process under standard conditions, the value of ΔA_{272} is calculated, and then using the simple relationships, the total concentration of the resulting chloroorganic compounds, TOX, can be estimated.

It was found that the value of ΔA_{272} is linearly correlated with the concentration of TOX in a sample of chlorinated water:

$$TOX_{powstałe} = k\,\Delta A_{272} \quad (1)$$

Korshin et al. [8] determined from a series of studies the correlation between the value of A_{272} and the TOX concentration determining that the value of the coefficient factor k is 10800 (μg×cm)/L. It has been suggested that the use of the ΔA_{272} measurement is an excellent indicator of the formation of general halogenated organic compounds (TOX), independent of chlorine to dissolved organic carbon (RWO) ratio, bromide concentration, pH, reaction time and NOM origin.

* Corresponding author: andrzej.jodlowski@p.lodz.pl

The purpose of the study was to determine the effect of removal of natural organic admixtures from surface waters in coagulation and in comparison coagulation and adsorption processes using PAC to reduce the TOX concentration resulting from chlorination and determined by differential absorbance.

2 Materials and methods

2.1 Water used during the study

The study was conducted using tap water contaminated with humus extract and using natural water taken from the Pilica River. Humus extract was prepared from garden peat, whose aqueous suspension was basified with sodium hydroxide to a pH of 10.0, and then filtered twice through a paper filter. The extract was added to tap water in the amount of 5 mL/L. Water from the Pilica River was collected during the flood from the chambers of intake of the water treatment plant (WTP) "Tomaszów". The results of the analysis of raw water prepared in the laboratory and the water from the Pilica River are presented in Table 1.

Table 1. Characteristics of waters used during the study.

Parameter	Tap water contaminated with humic extract	Water from Pilica River (filtrated sample)
Color, g Pt/m^3	117 - 154	176 -198
Turbidity, NTU	0.43 - 1,23	3.1 – 3.9
Temperature, °C	20.0 - 22,5	17.0 – 17.5
pH, -	7.45 – 8.45	7.64 – 7.69
Alkalinity, g $CaCO_3$/m^3	175	80 - 90
DOC, g C/m^3	7.39 – 10.58	21.7 – 25.3
Absorbance UV_{254}, m^{-1}	22.9 – 30.1	66.4 – 79.3
Absorbance UV_{272}, m^{-1}	23.4 – 28.0	50.4 – 65.0
SUVA, m^3/(g C×m)	2.38 – 4.07	3.06 – 3.13

2.2 Research on water purification

Standardized jar test procedure was used in the water purification studies. Coagulation was carried out with aluminum sulfate ($Al_2(SO_4)_3 \times 18\ H_2O$), which was dosed as 1% solution. The pH was adjusted during tests to the value 6.0. Rapid mixing time was about 3 minutes, slow mixing was carried out for 15 minutes, and sedimentation time was set for 30 minutes. Some of the experiments were carried out with the addition of powdered activated carbon (PAC) CWZ-22 type. According to the manufacturer, this PAC has a specific surface area of 800 - 950 m^2/g, and the pore volume is between 1.5 and 2.0 cm^3/g. PAC was added as an aqueous slurry during rapid mixing. Water quality was measured in raw water, water purified during coagulation and water after coagulation/adsorption after filtration through Munktell paper filter. Water, pH, color, DOC and absorption at 254 nm and 272 nm were measured in water under the investigation.

Chlorination of raw and purified water was carried out with a 1.25 mg Cl_2/L sodium hypochlorite solution after adjusting the pH value of 8.0. The mass ratio Cl_2: DOC = 1:1 was preserved. Absorbance at 272 nm was determined in water before chlorination and in chlorinated water and then the difference in these values (ΔA_{272}).

2.3 Applied analytical procedures

The water color and absorbance of UV254 and UV272 were determined using the HITACHI UV-VIS U-2001 spectrophotometer during the analytical inspection of water. UV-Spectra software was also used. Turbidity was quantified using a HACH 2100 IS turbidimeter. The CP-505 ELMETRON digital pH meter was used for pH measurement. DOC measurements were performed using the HACH-LANGE IL 550 TOC-TN analyzer.

3 Results and discussion

Both laboratory and river water were characterized by large content of high molecular weight organic substances with aromatic structure and hydrophobic properties as shown by the SUVA value, which exceeds 3.0 L/(mg C×m). The content of organic impurities expressed by DOC concentration was significantly higher in river water than in model water prepared on the basis of humus extract. SUVA values were slightly higher for river water (Table 1). The dose of coagulant for effective water purification was under the effect not only by the dissolved organic matter content and chemical nature but also by the turbidity particles, especially in the case of river water.

The effectiveness of organic matter removal as a result of both water types coagulation using aluminum sulphate is shown in Fig. 1. The technological effect of water contaminated with humic extract was obtained with significantly lower doses of coagulant than in the case of river water. It should also be noted that high molecular weight humic substances with double bonds were relatively easily removed, as evidenced by changes in ultraviolet light absorption at 254 nm (Fig. 1a). It can be concluded that the low molecular weight content that is not absorbed at 254 nm was relatively small. The course of coagulation of water from the Pilica River showed a more diverse composition of organic matter from the point of view of their fractional composition (Fig. 1b). It can be assumed that the concentration of hydrophilic low molecular weight substances was higher in river water. With regard to both types of water, a high degree of color reduction was obtained as a result of coagulation. For water with humus extract at 120 mg/L

https://doi.org/10.1051/e3sconf/20185900030

aluminum sulphate, the water color was reduced to 21 mg Pt/L, and the color of water coming from the Pilica River was reduced to 24 mg Pt/L at a coagulant dose of 150 mg/L.

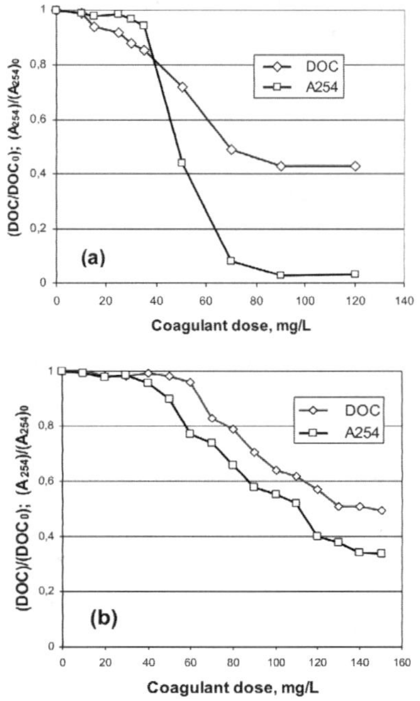

Fig. 1. Changes of DOC concentration and A254 absorption as a result of alum coagulation. a) tap water contaminated with humic substances, b) Pilica River water.

To enhance the removal efficiency of organic matter, especially in the case of low molecular weight fractions of hydrophilic properties, i.e. non-absorbing molecules at 254 nm, PAC in the dose of 10 mg/L was introduced to the purified water in addition to the coagulant. Changes in DOC level and SUVA values resulting from purification of water contaminated with humic extract are presented in Table 2, and changes in the values of the same indicators occurring during purification of water from the Pilica River are presented in Table 3.

Samples of raw and purified water were chlorinated under predefined conditions (pH of 8.0, Cl_2:RWO = 1:1, contact time = 24 h). The concentration of TOX was calculated based on the values of ΔA_{272} (Tables 2 and 3). It can be seen that the concentration of TOX generated was reduced with decreasing concentration of organic substances expressed by DOC values. Fig. 2 shows the dependence of the TOX concentration generated under the applied chlorination conditions of water versus the DOC concentration. It can be seen that the concentration of TOX generated was reduced according to decreasing concentration of organic substances expressed by DOC values.

The highest values of TOX were found out in the water sample taken from Pilica river during heavy rains. That kind of water was characterized by high DOC level. Treatment of water contaminated with humic extract as well as water from Pilica river by using aluminum sulfate coagulation, and especially after the introduction of PAC, have reduced the amount of organic matter and consequently reduced the amount of TOX resulting from chlorination.

Table 2. Changes of DOC, SUVA values and TOX concentration as a result of water containing humic substances treatment.

Water treatment	DOC, mg C/L	SUVA, L/(mg C×m)	ΔUV_{272}	TOX, µg/L
Raw water	10.56	2.54	0.094	1015
Coagulant dose (50 g/m^3)	6.23	1.76	0.046	497
Coagulant dose (70 g/m^3)	5,17	1.53	0.039	421
Coagulant dose (70 mg/L), PAC dose (10 mg/L)	4.97	0.92	0.010	108

Table 3. Changes of DOC, SUVA values and TOX concentration as a result of Pilica River water treatment.

Water treatment	DOC, mg C/L	SUVA, L/(mg C×m)	ΔUV_{272}	TOX, µg/L
Raw water	21.10	2.32	0.323	3488
Coagulant dose (50 g/m^3)	16.82	2.05	0.145	1582
Coagulant dose (70 g/m^3)	9.71	1.85	0.061	659
Coagulant dose (70 mg/L), PAC dose (10 mg/L)	3.31	0.54	0.017	186

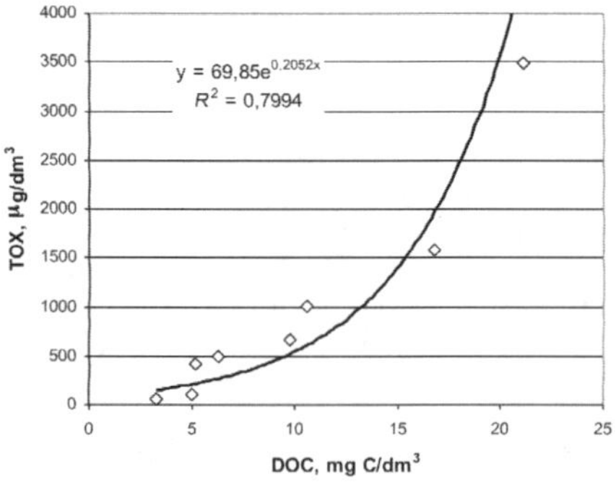

Fig. 2. Formation of total organic halogen (TOX) as a function of dissolved organic carbon (DOC).

The exponential relationship between the TOX formation potential and the DOC concentration could be also obtained based on the data presented in [6]. It includes, among others, raw water quality data of 27 samples taken at various US water plants. The water samples used in the studies were characterized by determining the alkalinity, TOC, DOC, UV254, SUVA, pH, turbidity and hardness. Chlorination was carried out using sodium hypochlorite under similar conditions as used in this study. The contact time was 24 h, the pH was 8.0, the temperature was 20°C. A slightly different dosage of chlorine was used as it was set so that the residual chlorine concentration was about 1 mg Cl_2/L. The TOX concentration was determined using the method described in Standard Methods [5]. The resulting relationship between the TOX formation potential and the DOC concentration in the raw water used is shown in Fig. 3.

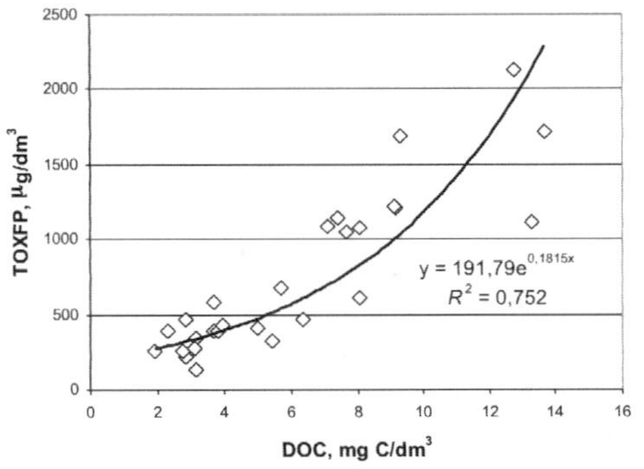

Fig. 3. Relation between TOX formation potential and DOC concentration in raw water from 27 US water treatment plants based on data taken from [6].

Differential absorbance also makes it possible to estimate the concentration of individual organochlorine compounds resulting from NOM chlorination [9].

DBPs concentration depending on $-\Delta A_{272}$ can almost always be presented in the form of linear equations with values of $R^2 > 0.95$. Korshin et al. [7] found that the only DBPs which did not agree with this trend was CH, which concentration correlated better with $-\Delta A_{272}$ according to the exponential relationship. TCP and DCAN were unstable at pH 7 and 8, but at pH 6, the linear relationship between their concentration and $-\Delta A_{272}$ was as strong as for those more stable DBPs. Dependence of DBPs concentration from $-\Delta A_{272}$ may be of practical importance as they provide an alternative approach to monitoring the emergence of individual DBP forms in on-line measurements.

Özdemir et al. [10] performed chlorination of raw water samples from Lake Terkos and Lake Büyükçekmece, which are sources of water supply for the city of Istanbul (Turkey). It was found that the spectral characteristics of the chromophore structures absorbed in the UV range by the action of chlorine were obtained. A strong linear correlation was found between ΔUV_{272} and TTHM ($R^2 > 0.96$). It has been observed that the group of THM compounds formed by the chlorination of raw lake water was dominated by $CHCl_3$.

4 Conclusions

1. As a result of coagulation of model water contaminated with humic extract and natural water taken from the Pilica River, the removal rate of dissolved organic substances characterized by DOC was 51% and 54%, respectively.
2. The use of PAC at the coagulation step has increased the removal of organic matter measured as TOC from water contaminated with humic extract to 53% and from river water to 84%.
3. Increased efficiency of organic matter removal from water by using coagulation, filtration and adsorption processes made it possible to reduce the TOX concentration by 89% in the case of tap water contaminated with humic extract and by 95% in the case of river water.
4. Differential absorbance (ΔUV_{272}) is an alternative approach to the estimation of TOX and THM concentrations, including chloroform level, formed by the NOM-containing waters.
5. Differential absorbance provides the ability to continuously monitor the concentration of halogenated organics resulting from the chlorination of water containing NOM.

References

1. P.C. Singer, S.D. Chang, J. AWWA **81,** 8 (1989)
2. H. Pourmoghaddas, A.A. Stevens, Wat. Res. **29**, 9 (1995)
3. T. Karanfil, M.A. Schlauntman, I. Erdogan, J. AWWA **94**, 12 (2002)
4. I.N. Najm, V.L. Snoeyink, B.W. Lykins, J.Q. Adams; J. AWWA **83,** 1 (1991)
5. *Standard Methods. For The Examination of Water and Wastewater,* 20th ed. (APHA, AWWA, WEF. Washington DC, 1998)
6. A.D. Archer, P.S. Singer, J. AWWA **98**, 8 (2006)
7. G.V. Korshin, M.B. Benjamin, Wat. Res. **31,** 4 (1997)
8. G.V. Korshin, M.B. Benjamin, Wat. Res. **31,** 7 (1997)
9. G.V. Korshin, M.B. Benjamin, O. Hemingway, W. Wu, *Development of differential UV spectroscopy for DBP monitoring* (AWWA Research Foundation; American Water Works Association, 2002)
10. K. Özdemir, I. Toröz, V. Uyak, TSWJ. Article ID 890854 (2013)

AUTHOR INDEX

EDP Sciences – Web of Conferences
17, Avenue du Hoggar
Parc d'Activité de Courtabouf
BP 112
F-91944 Les Ulis Cedex A
France

ISBN 978-1-5108-7316-2

9 781510 873162

CURRAN ASSOCIATES INC.
proceedings.com